Dictionary of Automotive Engineering

Second Edition

Don Goodsell
CEng, MIMechE, MSAE

Society of Automotive Engineers, Inc.
Warrendale, Pa.

Butterworth-Heinemann
Oxford London Boston Munich New Delhi
Singapore Sydney Tokyo Toronto Wellington

First published 1989
Paperback edition 1991
Second edition 1995

Printed and bound in the United States of America.

Butterworth-Heinemann Ltd
Linacre House, Jordan Hill, Oxford OX2 8DP

A member of the Reed Elsevier plc group

British Library Cataloguing in Publication Data
A catalogue record for this book is available from the British Library

Butterworth-Heinemann ISBN 0 7506 2795 6

Society of Automotive Engineers, Inc.
400 Commonwealth Drive, Warrendale, PA 15096-0001
Phone: (412) 776-4841 Fax: (412) 776-5760

Library of Congress Cataloging-in-Publication Data

Goodsell, Don.
Dictionary of automotive engineering / Don Goodsell. -- 2nd ed.
p. cm.
ISBN 1-56091-683-4 (hardcover)
1. Automobiles--Dictionaries. I. Title.
TL9.G64 1995
629.2'03--dc20 95-32289
CIP

SAE ISBN 1-56091-683-4

Cover photo reproduced with permission of Chrysler Corporation (1995 Dodge Avenger).

SAE Order No. R-159

To the memory of my cat Percy

Preface to First Edition

English is a living language, and one with an integrity worth preserving. Its rich store of words has been lifted over many centuries from many different tongues, and grows every year. The motor vehicle has been responsible for no mean contribution to that growth.

Of the more than two thousand entries in the present work, many are receiving their first published definition; others have become part of everyday communication. Some reveal the country of origin, as the French *carburateur*. Some, like aerodynamics, derive from classical sources. Many can be traced back to an earlier craft or trade, while the numerous eponyms take the name of an inventor or manufacturer whose identity may become lost in the passage of time—terms as old as the *Hooke's joint* or as recent as the *MacPherson strut*. Some are coined by the engineer, some by the salesperson, and some by the legislator. They all go into the vernacular melting pot and emerge as part of a living language to which the truck driver is as entitled to contribute as is the scientist.

Had this dictionary been intended for one isolated national market such as the US or the UK, the thorny problems of spelling and terminology would not have arisen. Spelling in particular is a subject that seems to arouse passionate national feelings, though more often of prejudice than scholarship.

The very scope of this work calls for an open-minded approach, a policy stemming from a desire to ease comunication. Faced with a choice of spelling, I have tried to choose the etymologically more correct, and then the simpler. Thus *tire* rather than *tyre*, and *balk ring* rather than *baulk ring*, both of which are supported by the Oxford English Dictionary, the Concise Oxford listing *tyre* among words "...in which -y- has intruded itself without dispossessing a more correct -i-." The most traditional of English gentlemen ties his tie, and *balk*, just for interest, derives from the Old Norse, the -u- being a spurious later addition, as incidentally was the -h- in Thames. *Color* appears without its -u- not because Americans spell it that way but because it is simply the old Latin word *color* which picked up two -u's- en route through France and only lost one on arrival. Thus what some readers may interpret as a preference for American English is merely an attempt to promote the fairer choice. It has nothing to do with nationalism.

A studious endeavour has been made to include words from all possible sources. To readers from English-speaking lands other than America and Britain this may appear as an arrogant indifference to their own vernacular, but it is merely ignorance. A letter to the publisher listing and defining such omissions for a subsequent edition will do far more than wrath to serve future readers.

Even if common sense does not argue the case for a unified technical language, the computer certainly does. The searching of databases becomes less effective and more costly the more variants there are of the target words. *Inlet, intake, induction* and *suction* all mean the same thing in an engine's cycle of operation, but only one may yield the required result.

The definition one gives to a word is determined by the person for whom it is defined. An explanation for a layperson may not satisfy an expert. This dictionary steers a middle course by assuming the knowledge required of the engineering student and acquired by many an enthusiast. It will therefore be within the grasp of all engineers and many others within the industry.

The words and terms selected are those that have an automotive connotation. General engineering terms are infrequently included as these can be found in works such as the *Dictionary of Mechanical Engineering* by G.H.F. Nayler, to which the *Dictionary of Automotive Engineering* is in many ways a companion—likewise the more esoteric aspects of fuels and lubricants, electricity and electronics.

Where the definition of a term may be made clearer by reference to another, the reader is invited to "See also" which terminates many definitions. Where a particular term is neither preferred nor in mainstream usage, the reader will be directed to another entry with "See..." for the full definition, for example, **blocking ring** See ***balk ring***. This does not imply official recommendation but merely reflects what is believed to be the trend in technical usage.

The definitions in this dictionary are intended to reflect usage, and not primarily to influence it. They do not carry the authority of official status (though some are derived from official documentation). Greater rigor is rendered impracticable by detailed but often significant differences from one country to another.

There is a simple lesson to be learned here for all who write technical material: if you are in the slightest doubt as to whether your readers will understand exactly what you mean, spell it out. You cannot assume that they have a copy of this dictionary by their sides (happy as I would be if they did) or of any official document.

Where national practice is deep-rooted and unlikely to change in the interests of conformity both words or terms are defined, with an indication of the country of usage, for example, **crossply (US: bias ply)**. The only exception to this rule occurs where such terms are similar, and would appear within a page of each other.

It would be naive to suggest that clear demarcations of terminology separate the US and Britain. For years I believed that after the Declaration of Independence the term kingpin had become taboo, and that was why Americans called it wrist-pin, but I was wrong. Trends in terminology are often more localized, expressing a manufacturer's preference, or revealing the influence of an earlier local industry such as shipbuilding or railway (US: railroad!) engineering.

Seemingly insoluble problems arise with the hyphen and capitalization. As a general rule, the hyphen is included where within technical text its omission might impair comprehension or cause ambiguity, or where two vowels might otherwise be juxtaposed. Three-way converter obviously needs its hyphen, as does pre-ignition, but there are many instances in which no hard-and-fast rule can be established. Within the dictionary hyphenated words have the same priority as would be accorded two words separated by a space, so that dual-drive takes its place between dual control and dual fuel, either of which might justifiably have carried a hyphen.

As with hyphenation, the capitalization of eponymous terms within the dictionary attempts to follow popular usage. The tendency has always been the loss of the capital as the term becomes more familiar. Younger generations often write diesel engine without the capital D, though it seems the MacPherson strut is assured of its upper case. Here, as elsewhere, an apparent lack of consistency merely reflects usage in the knowledge that some trends cannot be reversed.

Informal and slang terms are included because they are a part of the language and often, in time's fullness, are blessed with official sanction. The distinction between informal and slang is sometimes narrow, but it is made as a guidance to usage, particularly to those for whom English is a second language. The many omissions in such specialized pursuits as trials riding or drag racing stem from ignorance and too narrow a circle of friends. Genial enlightenment will be welcomed—we all have our blind spots.

Illustrations have been included where they are likely to clarify a meaning, consistent with limitations of space. Many of the drawings are of assemblies rather than individual items, and call up various components. Where possible these have been labeled to refer the reader to related entries, though it has not always been possible to give alternative names for parts so labeled. To help the reader who is referring to a component that is illustrated on an assembly drawing appearing under another, and often distant, entry, a figure reference will allow the assembly drawing to be located. The system is simple: *Figure S.5* means the fifth illustration in section "S"—an admittedly less exact but far more practical method than giving a page number.

That the *Dictionary of Automotive Engineering* should be published in America and Britain under a co-publication agreement between the Society of Automotive Engineers and Butterworth Scientific (now Butterworth-Heinemann) has made possible the production of a work that is international both in spirit and content. A national dictionary would have presented fewer problems, but would be less useful. While inevitably manufacturers follow national or company custom, the industry is international in outlook, and largely English speaking, even where English is not the first language.

Don Goodsell
1989

Acknowledgments

The following organizations have cooperated in providing material and advice. Their assistance is gratefully acknowledged:

Austin Rover Group Ltd
Automotive Products plc
Borg-Warner Ltd
Citröen UK Ltd
Clayton Dewandre
CMI-Tech Center Inc.
Corning Inc.
Crane Fruehauf Ltd
Cummins Engine Company Ltd
Dennis Specialist Vehicles
Eaton Corporation
Firestone UK Ltd
Ford Motor Co. Ltd
The Jacobs Manufacturing Company
Jost-Werke GmbH
Leyland DAF Ltd
Lucas Industries Ltd
Ringfeder GmbH
Road Transport Industry Training Board
Robert Bosch GmbH
Rockwell Automotive Body Components (UK) Ltd
Rubery Owen Holdings Ltd
Scania (Great Britain) Ltd
Society of Motor Manufacturers and Traders
Spicer Clutch Division UK (of Dana Corporation)
Telma Retarder Ltd
Textron Automotive Company
TI Automotive Ltd
TRW Cam Gears Ltd
VBG Produkter AB
J.M. Voith GmbH
Voith Turbo GmbH
Volvo Concessionaires Ltd
Volvo Trucks (Great Britain) Ltd

WABCO Westinghouse
Weber Concessionaires Ltd
Wellworthy Ltd
Jonas Woodhead and Sons plc
Zahnradfabrik Friedrichshafen AG

The following publishers or organizations have kindly given permission for the use of illustrations, some of which have been redrawn or annotated for use in this dictionary:

Butterworth-Heinemann, Oxford:

From *Motor Vehicle Mechanic's Textbook* (F.K. Sully): Figures A.2, B.2, B.4, C.14, F.4, H.1, H.3, M.2, O.2, S.7, T.2, T.11, W.4

From *Motor Vehicle Craft Studies* (F.K. Sully): Figures B.1, C.4, C.5, C.8, D.4, D.5, H.4, P.1, T.8, W.6, W.7

From *Vehicle Technology* (M.J. Nunney): Figures C.1, C.10, C.11, H.2, P.4, S.1

From *Aerodynamics of Road Vehicles* (W-H Hucho, co-published by SAE, translated from the original German edition published by Vogel-Verlag, Würzburg): Figure K.1

Edward Arnold, London:

From *Advanced Vehicle Technology* (Heinz Heisler, published by Edward Arnold: Figure A.5

The Editor, Classic Bike, Peterborough, UK:

From *The Motor Cycle*, June 1924 : Figure G.2

Centrex

Before the days of the computer the lexicographer worked with weighty files of cards, which had to be set in order and typed before they were submitted to the publisher. The word processor has taken much of the burden from that job, allowing definitions to be amended at random, automatically ordered, and presented on a disc hardly larger than a card from a card index. Because many users of this book will be the people behind the scenes who receive little recognition of their engineering ingenuity, I would like to pay credit to their counterparts who created the BBC microcomputers on which the text has been generated, to the software houses Computer Concepts, Norwich Computer Services and IFEL, and to the ever-helpful staff of the BBC user group, Beebug.

Many people have contributed something of their knowledge or experience to this dictionary, providing material, reading and checking definitions and offering advice. Among the many I would like to acknowledge the following by name:

Mr. Charles Beard, formerly of Ricardo and Co.; Mr. David Bell, mechanical engineer and expert on car restoration; Mr. Geoffrey Carr, Head of the Aerodynamics Department of the Motor Industry Research Association (MIRA); the staff of the motor vehicle section of Canterbury College, Kent, UK; the editor and staff of the magazine *Commercial Motor*; Mr. John Donald of MIRA; Mr. George Gaudaen, lexicographer and former standards engineer with the SAE; Mr. Leslie Lilly, formerly of Ricardo and Co., and editor of the much-praised *Diesel Engine Reference Book*; Dr. Peter Newcomb, author and lecturer in the Department of Transport Technology, Loughborough University; Mr. Keith Owen and Mr. Trevor Coley, fuels consultants; Mr. Michael Shields of MIRA, keen critic and man of wider literary endeavour; Dr. Robert Spurr of Ferodo and Co.; Mr. John Whitehead of MIRA.

The books, journals, standards, and papers used in preparing this dictionary are far too numerous to list. However, mention must be made of the Society of Automotive Engineers, and American Technical Publishers, who have been unfailingly generous in the provision of technical material, the library of the Institution of Mechanical Engineers in London for use of many facilities, and Mr. Michael Shields for selflessly putting at my disposal the information he had collected for a similar, but thwarted, project.

Preface to Second Edition

My aim in producing this new edition has been to extend coverage into areas that were only lightly covered in the first edition, and to add terms that have subsequently come into circulation. I have also extended the system of cross-referencing that readers seemed to find so useful in the first edition. The areas now substantially expanded include fuels and lubricants, materials (including plastics and elastomers), tires, off-highway and construction vehicles, testing, and electronics. I have tried to do this while observing the original concept, for which please see the Preface to First Edition. This is not, I repeat (and reassure), a book of standard, official definitions. It is about how words are used in the English-speaking world, as far as I can tell. It is a guide to understanding. I hope it will find you the word to use, and the meaning of a word you encounter.

Certain terms are subject to the (often changing) contraints of legislation, such as Gross Train Weight, Heavy Locomotive, where a current definition incorporates a weight limitation or some other factors that change with changes in national or international legislation. I have tried to imply the variability of such definitions and so bridge the gap between the workings of the legislative mind and the needs of vernacular communication. Outside the laboratories, design offices, and committee rooms of this hard-edged industry, millions of people talk about cars, trucks, and the vehicles with and in which they make their routine journeys, take their leisure, pursue their trades. Language, even technical language, belongs to all of us, and if I ask one thing of the professional reader, it is to accept this point. It is the point that justifies this dictionary.

It is bigger, but it remains essentially the same book as the first edition, with the same simple philosophy of definition. It is not an encyclopedia and is not intended as a stand-in for a textbook. People (individuals rather than libraries) buy dictionaries because they need them, or might one day need them, and those people may be among the many who, in their homes, speak an entirely different language while accepting English as the working *lingua franca* of engineering. Don't therefore be tempted to say, when finding a definition you consider commonplace, "But everyone knows that." Not, at least, until you have seen the world's population statistics of those who master the skill of communicating in English so that your lives and mine are made easier, and do so as well as mastering the skills of automobile engineering.

For the second edition, in addition to some of those previously listed who continued to offer help, I'd like to thank: Janet Ash, on chemical terminology; Robert Riley, whose delightful book *Alternative Cars in the 21st Century* I have come to know rather well; Arthur Caines and Roger Haycock, authors of *Automotive Lubricants Reference Book* for many useful definitions; staff of Michelin, with a European view of tire (tyre) terminology; Walter Bergman, whose efforts on behalf of the SAE's Tire Terminology committee have been prodigious; Graham Montgomerie, Chief Mechanical Engineer of the UK Freight Transport Association; Allan Lupton, lately of British Aerospace and classis car enthusiast for at least the 35 years I have known him; Alan Bunting, freelance commercial vehicle journalist for 25 years; Penny Bell for one delightful term I would otherwise have missed; *Classic Motor Cycle* magazine for stirring the memory of parts of bikes that were not classic when I rode them; J.C. Bamford for details of machines that, in the UK at least, are known, simply, and by almost everyone as JCBs; Aston Martin for some up-market suspension details; Reynard, competition car experts; Centrex, the UK truck training organisation; and although also mentioned with respect to the first edition, Keith Owen, who stoically undertook to read and correct a substantial set of definitions at a time when most others would have been celebrating an event far more important than any dictionary.

Don Goodsell
1995

A

A-arm Suspension frame consisting of two ***radius rods*** joined in the shape of an A, for providing lateral location. A ***wishbone*** (UK). Also ***A-frame***.

A-frame (1) Structural frame in the shape of an A. (2) A-shaped towing frame by which a truck is coupled to a ***drawbar trailer***. (3) Suspension frame in the shape of an A. See ***A-arm***.

A-pillar See ***A-post***.

A-post Structural member forming the forward corner of the cab or passenger compartment. In structural analysis, the post may be assumed to include adjacent (contingent) parts of the door frame. Also ***A-pillar***. See Figure B.4.

ABS See ***anti-lock brake system; acrylonitrile-butadiene-styrene***.

ABS relay valve Electrically actuated pneumatic valve that controls the air pressure in the brakes of a pneumatic ***anti-lock braking system*** during braking.

absolute viscosity See ***dynamic viscosity***.

absorption In acoustics, the reduction of noise by the application of a material that absorbs sound energy in the appropriate frequency range. See also ***cancellation***.

accelerated hardening Hardening, for instance of an adhesive, sealant, molded component, or finish, by the application of heat or catalytic action, or both.

accelerated weathering test Laboratory test on materials, components or complete vehicle that simulates the effects of weather on an accelerated time scale. See also ***salt spray test***.

accelerator Pedal by which the fuel flow to the engine is controlled, depression of the pedal causing the vehicle to accelerate.

accelerator heel point Assumed point of contact of driver's heel with floor when foot is placed on the undepressed accelerator. Also ***AHP***.

accelerator pump Carburetion system that provides ***enrichment*** to offset mixture weakening when the throttle is opened rapidly on acceleration.

accumulator (1) Rechargeable electrical storage battery. (Obsolescent) (2) A hydraulic accumulator for storing hydraulic pressure.

Ackermann steering System of ***double-pivot steering*** in which two steered wheels pivot about a vertical axis and are steered by linked ***steering arms***. The system was devised by Lankensperger but takes its name from the patent agent Ackermann. It was originally introduced to prevent capsizing of horsedrawn vehicles when turning sharply. See Figure A.1. See also ***Jeantaud steering***; ***reverse Ackermann***.

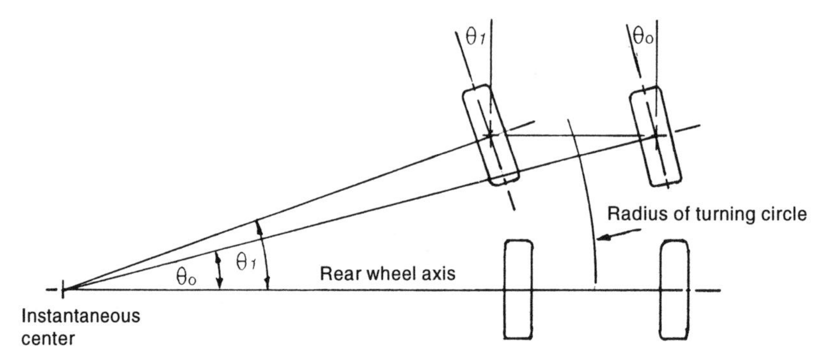

Figure A.1 Idealised Ackermann steering geometry, in which the projected axes of steered (front) wheels and of rear wheels coincide. (Track exaggerated for clarity.)

acoustic impedance Measure of the ability of a material to transmit sound. See also ***absorption***.

acrylic (1) Of the unsaturated acrylic acid or its compounds. (2) An acrylic resin-based substance, such as a paint or adhesive.

acrylic resins Resins based on acrylic acid, ***monomeric*** or ***polymeric***, including methyl methacrylate, which range from fluid to solid. As molding materials noted for their transparency and used as an alternative to glass for curved canopies, also in colored or opaque sheet form, and as the basis of paints. Some formulations have high impact toughness.

acrylonitrile-butadiene-styrene ***Polymeric*** material used mainly in injection molding and extrusion, and in sheet form. Of good tensile and impact strength, and heat resistance. Also ***ABS***.

active restraint Occupant restraint system requiring manipulation by wearer, as for example a ***seat belt***.

acute toxicity Toxic effect resulting from a single exposure to a toxic substance. See ***chronic toxicity***.

additive A small amount of chemical substance added to improve the properties of a fluid such as a ***lubricant*** or ***fuel***. See also ***detergent***, ***dispersant***.

adhesion (1) Grip between road and ***tire***, proportional to the static coefficient of friction. (2) Joining of two surfaces by bonding using an intermediate agent called an ***adhesive***. (3) Measure of the strength of a bonded joint, as for example in peel or in tension.

adhesive Material, usually in ***thixotropic*** or liquid form, for joining two surfaces by chemical or physical change or other process.

adiabatic engine Engine in which combustion heat loss to coolant is minimized. Also ***low heat rejection engine***.

admission stroke See ***induction stroke***.

adsorbent storage Form of low-pressure storage of certain gaseous fuels, such as ***methane***, through the adsorbent properties of activated carbon or other agencies. See also ***adsorption***.

adsorber coated substrate Adsorbent material, particularly one employed to temporarily trap emission gases prior to ***light-off*** of a ***catalytic converter***.

adsorption Adhesion of a substance to a surface, as with some ***lubricant additives*** to metals. See also ***desorption***.

advance stop Mechanical contact to restrict ignition advance in a ***vacuum advance*** system. See also ***retard stop***.

advanced ignition Maladjustment of ignition timing in a ***spark-ignition engine*** whereby the spark occurs before its optimum setting.

aerial (US: antenna) Wire or rod, often retractable, for receiving radio signals.

aerodynamic noise Noise generated by the flow of air around and through a vehicle due to its forward motion, excluding noise of ***forced ventilation***.

aerodynamic stability Response of a moving vehicle to air perturbations such as gusts, side winds and disturbances from passing vehicles.

aerofoil (US: airfoil) Streamlined planar shape such as a wing intended to produce positive or negative lift or otherwise derive some effect from the flow of air. See also ***Gurney flap***; ***wing***.

aerosol Finely divided dispersion of a liquid or solid in a gas.

afterburner Device that completes the combustion of incompletely burned exhaust products within the ***exhaust*** system. See ***emission control***.

aftercooler Heat exchanger that cools the ***induction*** air before it enters the cylinders of an engine, most often used in diesel engines downstream of a ***turbocharger***. Often (and incorrectly) called an intercooler.

after-running See ***run-on***.

agglomerator Separator or trap for removing water from fuel or oil. See also ***filter***; ***sedimenter***; ***separator***.

aggregate body Commercial vehicle truck body for the conveyance of aggregates, ballast and similar bulk materials, usually equipped with tipping gear.

agricultural commodity trailer Off-highway trailer for conveyance of agricultural produce and bulk material.

agricultural tractor On- and off-highway towing vehicle designed for the towing of agricultural equipment and trailers, and for general work on the land, and usually fitted with large-diameter wheels or a track laying system.

aiming Adjustment of direction of beam of ***lamps*** to meet highway or legislative requirements.

aiming screws Screws for adjusting the aiming of ***headlamps***.

air-assisted hydraulic brake Commercial vehicle braking system in which the ***master cylinder*** of a hydraulic braking system is actuated by pneumatic pressure or vacuum. Also ***air-over-hydraulic brake***; ***vacuum-assisted hydraulic brake***; ***vacuum-over-hydraulic brake***.

air bag Passive vehicle occupant ***restraint system*** in which on impact the rapid inflation of a cushion fitted in the ***fascia*** region restrains the body above torso level and prevents direct contact with the vehicle structure.

air bellows Usually cylindrical or torose rubber vessel or sleeve which, when filled with a compressed gas or air, acts as a compression spring. Mainly used in commercial vehicle or trailer suspensions. See also ***air spring***; ***air suspension***; ***gas spring***.

air bleed passage (UK: compensating jet) Jet or passage in a ***carburetor*** with a branch to atmosphere through which air is drawn and introduced into the fuel flow at high flow rates, thus preventing over-richness.

air brake (1) Brake in which the force that actuates the brake mechanism is provided by compressed air acting on a diaphragm within a brake chamber or ***servo***. Used mainly in commercial vehicles. (2) Aerodynamic ***spoiler*** for retarding high-speed vehicles.

air cell A small chamber, located in the cylinder head particularly of an ***indirect injection engine***, in which combustion is initiated. The burning fuel charge passes through a ***throat*** to the main ***combustion chamber***. Also ***swirl chamber***. See also ***antechamber***; ***Comet head***; ***indirect injection***; ***Lanova***; ***pre-chamber***.

air check valve One-way valve in ***pulse air injection*** system to prevent return of ***exhaust gases*** to induction system.

air cleaner Any device such as a porous paper, wire mesh filter or oil-bath cleaner, that prevents airborne particles from entering air-breathing machinery. Also ***air filter***.

air compressor Machine for delivering compressed air, as for example to an ***air brake system***.

air-cooled Cooled by the passage of air as opposed to water, etc. Describes engines in which combustion heat is lost mainly through radiation and convection from ***finning***. See Figure A.2.

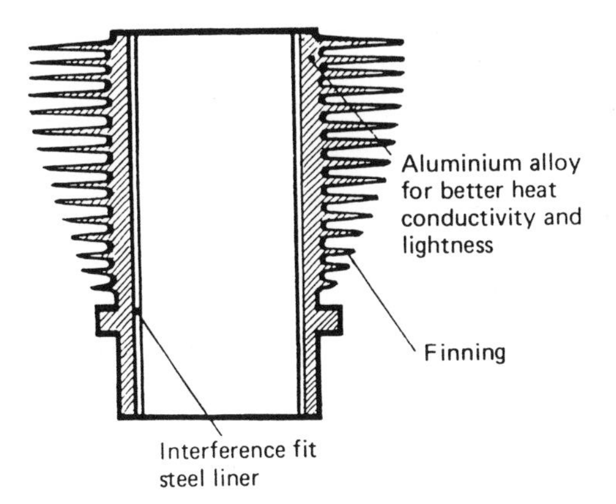

Figure A.2 Cylinder of air-cooled engine showing finning and liner (sleeve).

air dam Aerodynamically shaped transverse extension or spoiler below front fender/bumper which reduces drag created by vehicle underbody. Also ***apron***; ***underbumper apron***. See also ***spoiler***.

air deflector Contoured panel mounted on the roof of a commercial vehicle cab to improve air flow between ***cab*** and body.

air dryer Device for removing moisture from air, but particularly from the compressed air of a commercial vehicle braking system.

air filter See ***air cleaner***. Air filter is the customary UK term.

air-fuel delivery ratio Mass ratio of air to fuel inducted by an engine. The ***mixture strength.*** (Informal) See also ***stoichiometric ratio***.

air horn (1) Audible warning device in which sound is produced by a blast of air through a reed or resonator. (2) Engine induction tract leading from air filter or cleaner to a ***carburetor*** or inlet manifold. (US informal)

air-injection Addition of an air stream under pressure, but particularly to an ***exhaust system*** to promote combustion of ***unburned hydrocarbons*** and conversion of ***carbon monoxide*** to ***carbon dioxide***.

air-lift axle Pneumatically operated lifting axle of ***tandem axle*** commercial vehicle undercarriage. In some examples lifting or lowering are achieved by manipulation of pressure in the air bellows of an air suspension.

air-line Pressure or vacuum-resisting tubing or piping for connecting the components of a pneumatically operated system.

air-line connector Plug and socket for making connections within a pneumatic system. See also ***gladhand***; ***suzies***.

air-management kit Combined items for improving the aerodynamic efficiency of a vehicle, and particularly a truck, comprising for example ***air dam***, ***air deflector*** and other devices

air-over-hydraulic brake Hydraulic brake system actuated by an air-hydraulic power unit, mainly used on commercial vehicles. Also ***air-assisted hydraulic brake***.

air reservoir Pressure vessel for retaining compressed air, particularly for the braking system of a heavy vehicle.

air resistance See ***drag***.

air scoop Normally forward-facing raised aperture on vehicle bodywork to act as an intake for ***ram air***, as for example for ***engine cooling*** or ventilation. See also ***NACA duct***.

air seat Pneumatically suspended driver or passenger seat, particularly as fitted to a commercial vehicle, military vehicle, or other vehicle where ride quality might otherwise be unacceptable.

air shield (US: air deflector) Flat or contoured plate normally extending vertically from the cab roof of a commercial vehicle or any vehicle towing a substantially taller trailer, to improve airflow. Also ***dragfoiler***.

air shift PTO Pneumatically engaged ***power take-off***.

air silencer Device placed at the entry to an ***induction system*** to attenuate the noise of ***induction***.

air spring Spring using the compressibility of air or other gas to react against the imposed load. See also ***air bellows***; ***air suspension***; ***gas spring***.

air strainer An ***air filter*** or air cleaner. (US informal)

air suspension Vehicle suspension in which air in compression is the main or only spring medium. Also ***pneumatic suspension***. See also ***air bellows***; ***air spring***; ***Hydragas***; ***Moulton suspension***.

airless injection Fuel injection by mechanical pressure only. The normal method of injection in a ***diesel engine***. Also called ***solid injection***.

airscoop See ***scoop***.

alcohol Organic compound containing a hydroxile group, the basis of many synthetic ***solvents***, and, of ***ethyl alcohol***, an ***alternative fuel***.

alternator Alternating current electrical generator.

aliphatic Hydrocarbon compound with open carbon chains rather than ring structure. See also ***aromatic***.

alternative fuel General term for any automotive fuel other than the proprietary pump fuels ***gasoline*** (petrol) and ***diesel fuel***. See also ***alcohol***; ***CNG***; ***hydrogen***; ***LPG***; ***LNG***.

aluminum (UK: aluminium) Soft, ductile element of high electrical conductivity. For engineering structural purposes it is used in alloy form, alloyed with copper, manganese, zinc, silicon and other elements depending on properties required. Resistance to atmospheric corrosion varies with alloy composition. Extensively used in commercial and public service vehicle bodywork. Available in wrought and cast forms, and extrusions.

aluminum/air battery Battery using ***aluminum*** anode and atmospheric oxygen cathode, the aluminum being converted to aluminum hydroxide in the process. The battery is not rechargeable.

ambulance Vehicle designed and manufactured specifically for conveying the sick, injured or disabled.

anaerobic Without air, in the sense of a process that takes place with the exclusion of air.

anaerobic adhesives Wide range of adhesives formulations, many of which are based on ***acrylics***, which polymerise when oxygen is excluded. Applications include thread locking and locking of machine parts.

anchorage Point of a vehicle structure to which a non-structural stress-carrying item is attached, as for example a ***seat belt***, or seat.

anechoic Not reflecting sound, but totally absorbing sound. Literally, giving no echo.

anechoic chamber Acoustic test facility in which the inverse square law from a point source of sound applies, and where there is silence if no sound source is generated within.

angle of lock The horizontal angle between the plane of a ***steered wheel*** when cornering, and the plane when adjusted for straight ahead. See also ***lock***.

angledozer Bulldozer equipped with a blade set at an angle to the vehicle's direction of motion so that spoil is discharged to one side while the vehicle is in steady motion.

Annoyance Index A measure of acoustic annoyance, usually calculated as the sum of individual weighted noise parameters.

annulus (1) A ring-shaped component. (2) The annular internally toothed wheel of an ***epicyclic*** (planetary) geartrain. See Figure P.3.

antechamber See also ***air cell***; ***indirect injection***; ***prechamber***.

antenna (UK: aerial) Wire or rod, often retractable, for receiving radio signals.

anthropometric dummy Mannikin, or full-scale model representing the human form, for use in vehicle testing, such as impact testing, and in ***ergonomic*** design.

anthropometry Study of the measurement of the human body. See also ***ergonomics***.

anti-backfire valve Valve that allows air to flow from an air-pump into ***induction manifold*** on deceleration to prevent ***backfiring***.

anti-compounding valve Valve in air brake circuit that prevents driver from applying ***air brakes*** and ***spring brakes*** simultaneously, and so overloading the brake mechanism. Also called ***differential protection*** valve.

anti-dieseling valve See ***overrun fuel cutoff***.

anti-dive ***Suspension geometry*** that reduces or prevents nose-down pitching on braking.

anti-jackknife system Means of preventing a ***jackknife*** of an ***articulated vehicle***, usually by locking the tractive unit and ***semi-trailer*** through the ***fifth-wheel*** coupling.

anti-knock Counteracting a tendency to ***knock*** in an engine, as of fuel or combustion chamber design.

anti-lacerative glass Glass that does not splinter on fracture, and therefore diminishes the risk of laceration injury. See also ***safety glass***.

anti-lock braking system System that automatically controls wheel slip or prevents sustained wheel-locking on braking. Widely called by acronym ABS (regrettably the same as that of polymer acrylonitrile butadiene styrene). Also ***anti-skid*** (informal); ***wheel slip brake control system***. See also ***wheel speed sensor***.

anti-lock modulator Device that varies the braking force in accordance with the signal it receives.

Antiknock Index Specification of the ***anti-knock*** properties of a gasoline fuel, particularly in North America where it is defined by half the sum of the ***Research Octane*** and ***Motor Octane*** Numbers.

anti-oxidant Chemical that inhibits oxidation of materials with which it is in contact.

anti-roll bar Transverse torsion bar attached to a vehicle underside that couples the vertical displacements of ***nearside*** and ***offside*** wheels and in so doing reduces the vehicle's displacement in roll. Use of term ***roll-bar*** is discouraged as this is often used in a different context. Also ***stabilizer***. See also ***anti-sway bar***. See Figure A.3.

anti-run-on valve Valve which prevents ***running on*** or "dieseling" of a spark ignition engine when the ignition is switched off.

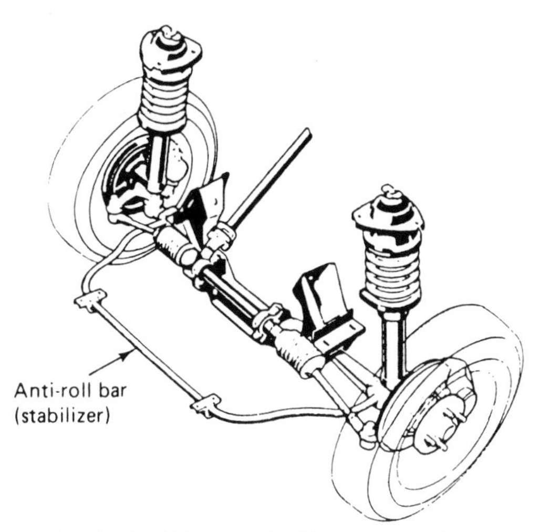

Figure A.3 Anti-roll bar on MacPherson strut front suspension.

anti-sail bar Horizontal bar to restrain deflection in service of an ***anti-spray flap*** or ***mudflap***.

anti-skid See ***anti-lock braking system***.

anti-spin regulation Control or prevention of ***wheelspin*** or ***wheelslip*** under power, normally by electronic sensing in conjunction with ***anti-lock braking***. Also ***ASR***.

anti-spray flap Flexible flap or curtain attached behind roadwheel to reduce road spray.

anti-squeal shims Shims inserted between brake piston and backplate to reduce brake ***squeal***.

anti-sway bar Suspension member, particularly applied to ***beam axle*** rear suspensions, that limits vehicle body lateral movement or sway.

anti-vibration mounting Flexible mounting, as for an engine or other mechanical item, that reduces the transmission of noise and vibration from the mounted item to a structure such as a vehicle chassis.

antifreeze A chemical, such as ***ethylene glycol***, added to the cooling water of an engine to depress the freezing point for winter operation.

antipercolator Tube and orifice in ***carburetor*** through which fuel vapor can escape from main jet tube to prevent over-enrichment due to vapor pressure.

apex seal Combustion gas seal of the ***Wankel engine***. See Figure W.1.

apparent viscosity Viscosity of a liquid measured at a given shear rate for cases where viscosity is shear-rate dependent.

appliance Specifically in automotive context, a fire fighting vehicle. Contracted from ***fire appliance***. (Mainly UK usage)

approach angle Maximum ***ramp angle*** that a vehicle can approach without fouling any part of the vehicle. See Figure L.3.

apron Downward panel extension, usually transversely mounted at the front of a vehicle, to reduce drag resulting from disorderly flow under the vehicle. Also ***air dam***; ***underbumper apron***. See also ***spoiler***.

aquaplaning (US: hydroplaning) Effect whereby a vehicle tire rides up on a thin surface of water and in so doing loses contact with the road surface resulting in sudden loss of ***traction*** and control.

aromatics Unsaturated hydrocarbon ***additives*** containing at least one benzene ring, such as toluene, xylene and related hydrocarbons, which may be added, as appropriate, to ***fuels*** and ***lubricants***.

arrest marks See ***beach marks***.

artic Articulated commercial vehicle ***tractor unit*** with ***semi-trailer***. (UK informal) See Figure A.4.

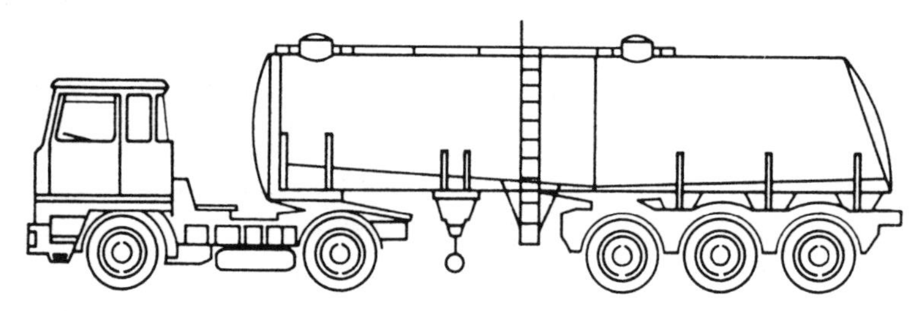

Figure A.4 An articulated road tanker.

articulated bus Usually single-decker bus with central articulation and accommodation for passengers in tractor and trailer units. See also ***close-coupled city bus***.

articulated vehicle Vehicle consisting of two or more usually separable wheeled units, such as a towing vehicle, or ***tractor***, and a ***semi-trailer***. Articulation is primarily in the steering mode, though some degree of horizontal axis articulation will be necessary to enable the vehicle to negotiate road surface irregularities. In official terminology usually a tractor and semi-trailer commercial vehicle combination with a small proportion of the trailer's load carried by the tractor. Also a ***close-coupled city bus***. See also ***artic***; ***articulated bus***; ***drawing vehicle***; ***truck***; ***van***.

ash Non-combustible residue of a fuel after combustion.

aspect ratio Ratio of length to width or, in the context of aerofoils, of span to average chord.

asperities Microscopic projections from a surface that are significant in wear and lubrication.

Aspin engine Four-stroke engine configuration incorporating a conical rotary valve which is itself part of the combustion chamber.

aspiration Breathing or induction process of an engine. Non-pressure-charged engines are often referred to as ***naturally aspirated*** or ***normally aspirated***.

ASR See ***anti-spin regulation***; ***traction control***.

Aston Martin linkage Form of straight-line suspension linkage introduced by maker of that name.

asymmetrical beam Light beam in which the light distribution is not symmetrical with respect to the median vertical plane of the beam.

atomization Conversion of a liquid into a spray of very fine droplets.

Austin Hayes transmission See ***Hayes transmission***.

auto-ignition Continued running of a spark ignition engine after the ignition has been switched off. Also ***dieseling***; ***running-on***.

auto transporter Truck and trailer combination for conveying vehicles, usually in service of the motor trade.

autocar (1) A passenger car. (Archaic) (2) A passenger coach. (Contemporary French usage) See also ***charabanc***.

autocycle Light ***motorcycle*** with pedals for assisting the motor on starting and on inclines. (Obsolescent) The forerunner of the ***moped***.

Autogas Liquefied petroleum gas motor fuel. Trade name in UK. Also ***LPG***.

autolube Pressurized oil metering system on a ***dry-sump*** engine, such as that of a four-stroke motorcycle.

automatic Vehicle with ***automatic transmission***. (Informal)

automatic braking Automatic brake application of a towed vehicle in the event of separation from the towing vehicle.

automatic choke Thermostatically operated choke valve in ***carburetor*** inlet tract, closed when engine is cold and opening automatically as engine or ***induction air*** temperature rises.

automatic gearbox (US: automatic transmission) Geared transmission unit in which gear ratios are automatically selected and engaged without the need for driver intervention. See also ***pre-selector***. See Figure A.5.

automatic leveling system Suspension system that automatically adjusts front and rear ***ride heights*** to compensate for changes in axle load.

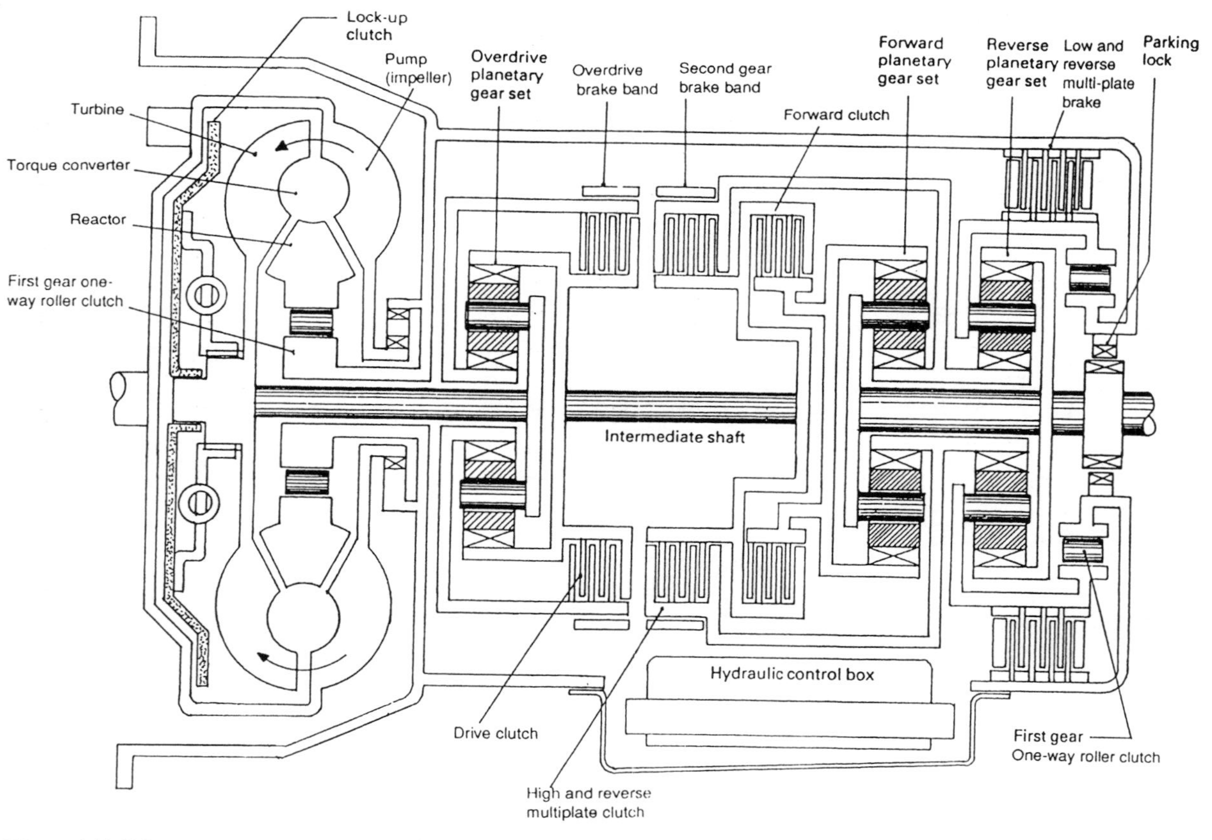

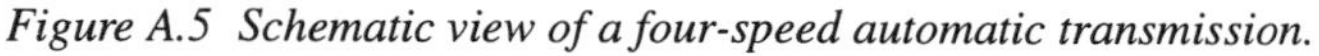

Figure A.5 Schematic view of a four-speed automatic transmission.

automatic slack adjuster (1) Device that automatically maintains the correct disposition of pushrod and cam lever in an air brake system. See also ***slack adjuster***. (2) Any device that compensates automatically for slackness in a mechanical system.

automatic speed control Device or system capable of maintaining selected vehicle speed in changing road conditions. See also ***cruise control***.

automatic transmission (1) Transmission system or gearbox in which gear ratios are selected and engaged automatically, though usually with provision for the driver to manually override the selection. (2) An automatic ***continuously variable transmission*** employing a variable-ratio friction drive such as a belt with expanding pulley. See also ***Hayes transmission***; ***pre-selector***; ***van Doorn transmission***. See Figure A.5.

automatic wear adjuster Device that automatically compensates an actuating system for wear in the item being operated, such as a clutch or brakes.

automobile (UK: passenger car) Self-propelled or motorized land vehicle, but particularly a private passenger car.

auxiliary brake Any brake that serves in addition to the main braking system of a vehicle. See also ***engine brake***; ***exhaust brake***; ***retarder***.

auxiliary driving lamp Lamp intended to provide illumination forward of the vehicle and to supplement the upper beam of a standard headlamp system. See also ***fog lamp***.

auxiliary gearbox Gearbox used in conjunction with a main change-speed gearbox to provide an extra range of speeds, as for example by providing a 2:1 reduction in addition to normal gearbox speeds. Also ***auxiliary transmission.*** (Mainly US) See also ***range-change***; ***splitter transmission***.

auxiliary lamps An official term describing ***fog lamps*** and ***spot lamps*** of a vehicle.

auxiliary transmission See ***auxiliary gearbox***.

auxiliary venturi See ***secondary venturi***.

average piston speed See ***mean piston speed***.

axle Horizontal transverse shaft or beam with spindles on which road wheels are mounted. See also ***dead axle***; ***fully floating axle***; ***half shaft***; ***live axle***; ***stub-axle***; ***transaxle***. See Figure D.5.

axle bearing Any bearing that supports an axle or ***half shaft***.

axle camber Wheel camber achieved in a beam ***dead axle*** by convex curvature of the axle.

axle carrier See ***axle casing***.

axle casing (US: axle housing) Rigid non-rotating casing that carries an axle or ***half shaft***.

axle chuckle Noise emanating from rear ***live axle***, usually due to excessive running clearances at the differential casing.

axle fore-and-aft shake Longitudinal oscillatory motion of an axle.

axle housing See ***axle casing***.

axle lift Mechanism for raising and lowering a ***lifting axle*** of a commercial vehicle.

axle ratio (1) Overall ratio between engine revolutions and revolutions at driven wheels. (2) The final drive ratio.

axle shaft See ***axle***.

axle side shake Lateral oscillatory motion of an axle.

axle sleeve Tubular casing of a live or dead axle shaft. See also ***axle casing***.

axle spindle Shaft machined to carry wheel bearings and seals and with means for securing the wheel to the axle.

axle spread Longitudinal distance, or effective "wheelbase" between the axles of a tandem- or tri-axle suspension unit. See Figure T.1.

axle tramp (1) Form of ***wheel hop*** on ***live axles*** in which the left- and right-hand wheels hop in opposite phase. (2) Resonant oscillation of an axle/suspension system, usually due to ***wind-up*** of an insufficiently damped suspension during heavy acceleration. The term is also used to describe a similar effect on vehicles without a ***beam axle***.

axle weight Static weight imposed on the highway by the wheels of one axle.

axle wind-down Reversal of ***axle wind-up***, effectively the rotation of the axle casing of a ***Hotchkiss*** type axle on braking or on going too suddenly into reverse. Often the term wind-up is used for both senses of rotation, i.e., for wind-up and wind-down. See Figure A.6.

axle wind-up (1) Torsional deflection of an axle shaft, as due to sudden application of power or brakes. (2) Rotation of an axle casing due to flexure of ***semi-elliptical springs*** in reacting torsional loads. (3) Oscillatory motion of an axle about the horizontal transverse axis through its center of gravity. See also ***axle tramp***. See Figure A.6.

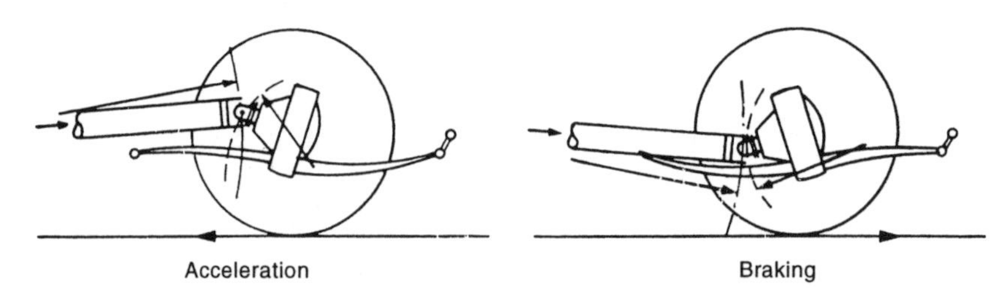

Figure A.6 Axle casing torque reaction and the need for a sliding joint and rear universal joint in the propeller shaft.

B

B-pillar See ***B-post***.

B-post Nominally vertical structural support of the roof of a vehicle, against which the front door closes. Also ***B-pillar***; ***central pillar***. See Figure B.4.

Babbitt metal An alloy of tin, copper, antimony and lead formerly used for engine bearings. Sometimes called Babbitt's metal.

back axle The rearmost axle (usually of a two-wheeled vehicle). Also ***rear axle***.

back-plate (1) Non-rotating plate carrying the shoes of a ***drum brake***. (2) Metal member carrying friction material of pad of ***disc brake***. See Figure D.8.

back pressure Pressure resisting the flow of a gas in a pipe, as in an ***exhaust pipe***.

back-up alarm (UK: reversing bleeper) Acoustic warning that automatically sounds when a vehicle, particularly a commercial or public service vehicle, engages reverse gear.

backbone chassis Chassis in which a single structural spine carries all powertrain and running gear, the body shell generally forming a secondary structure. See also ***punt-chassis***; ***spine-back***.

backfill (1) Of earth-moving equipment, to make good dug or uneven ground by filling from a loading shovel or other device, usually while retreating over the ground being prepared. (2) Material used in backfilling.

backfire An explosion of unburned or partially burned fuel in an exhaust or inlet system. See also ***anti-backfire valve***.

backhoe A mechanical digging bucket, usually hydraulically actuated, attached to the back of a vehicle, such as a ***backhoe loader***.

backhoe loader Earthmoving vehicle equipped with a loading shovel at the front and digging hoe at the back.

backlash Loss of motion between the input and output of a mechanical system, as due to looseness or flexure. Term implies some restoring torque at the input, though no effective output in the backlash phase, as for instance in a system of gears and/or linkages. See also ***play*** (which generally implies an insignificant restoring torque).

backlight The rear window or screen of a vehicle. (Slightly archaic)

backlight defogging system (UK: rear screen heater) System for clearing moisture from the interior surface of a rear window of a vehicle, either by an electrical element within the glass or heated air from a blower.

backrest Rear part of a seat against which the back rests. Also ***squab***.

backup lamp (1) **(UK: reversing lamp)** Lamp used to provide illumination behind a vehicle, particularly when reversing. (2) Lamp to supplement a standard headlamp system.

backup system Any system that, usually automatically, maintains a system in operation in the event of a major failure or breakdown of that system, as for example the pressure or vacuum of a braking system.

baffle plate An internal transverse plate in a ***muffler*** or ***silencer***, or in an oil ***sump*** or ***tank***.

balance beam Beam or lever that couples the suspensions of the two axles of a tandem rear axle arrangement of a heavy vehicle, thus making the suspension ***reactive***. On a leaf spring suspension a balance beam might couple the adjacent ***eye-ends*** of the leafs. See also ***reactive suspension***; ***walking beam***. See Figure T.1.

balance pipe Pipe or tube joining the venturis of twin carburetors.

balance shaft A rotating shaft which counters unbalance in a rotating machine, such as an engine, by means of an ***harmonic balancer*** or ***vibration damper***.

balanced crankshaft Engine ***crankshaft*** designed so that the disposition of its mass counteracts the out-of-balance effects of the crank and sometimes the reciprocating components. Balance is usually achieved by extending the crank webs to form counter-balances.

balk ring Rotating component of a ***synchromesh gearbox*** that prevents or "balks" premature engagement of gears. Sometimes baulk ring or ***blocking ring***. See Figure G.1.

ball and socket Mechanical joint in which a spherical end moves freely within a recessed socket. Used on suspension and steering linkages. Also ***ball joint***. (Informal)

ball bearing Rolling element bearing in which hardened balls run in tracks formed in inner and outer races. Various types exist, capable of carrying journal loads, thrust loads, or a combination of loads.

ball joint See ***ball and socket***.

ball thrust bearing ***Rolling element bearing*** for carrying axial loads by way of hardened steel balls running between races. Sometimes, though informally, called a ***thrust race***.

ballast resistor Electrical resistor used to regulate ***ignition coil*** output at higher engine speeds and increase spark voltage for cold starting.

balloon tire Low-pressure, bulbous sectioned tire used in the late 1920s and 1930s.

banana spring ***Coil spring*** with curved centerline axis.

band brake External band with friction lining, sometimes used as an additional parking brake, and as a change-speed brake in certain types of ***semi-automatic gearbox***.

banjo Hose fitting that connects a hose at right angles to the axis of a pipe. The circular body with one radial pipe connector gives the appearance of the musical instrument from which the item takes its name.

banjo axle Drive axle with a drum-shaped final drive or differential case. See also ***split axle casing***.

Barber-Greene Controllable ***tailgate*** on a tipper by which the rate of discharge of load, particularly road-making materials, can be regulated. (Informal)

barrel (1) Main air passage through a ***carburetor***, in which air and vaporised fuel are mixed. Also ***mixing chamber***. (2) The traditional unit of volumetric measure of petroleum, equivalent to 159 litres.

barrel cranking motor Starter or cranking motor for engines with small flywheels, the pinion being attached to a barrel to increase inertia.

barrel tappet A hollow cylindrical ***tappet***, sometimes with facility for adjustment.

barrel valve lifter A ***barrel tappet***. (Alternative US terminology)

barrier Massive or firmly located vertical obstruction for test evaluation of vehicle impact. See also ***moving barrier***.

base blend Primary liquid constituents of a ***fuel***, to which ***additives*** are added.

base stock Primary liquid constituent of a ***lubricant***, to which ***additives*** are added.

bath tub Combustion chamber shaped like an inverted bath tub in the base of which the valves are seated. See Figure C.14.

battery Direct current electrical storage unit which converts chemical or other forms of energy into electrical energy. Also ***accumulator***. (Archaic)

battery charger Static device for recharging a storage battery.

battery ignition Conventional ignition system using battery and coil, as opposed to a rotating magneto. Also ***coil ignition***; ***Kettering ignition***.

battery shedding Loss of material from a battery plate, as caused by age or repeated overcharging.

baulk ring See ***balk ring***.

Baverey compound jet Submerged ***carburetor compensating jet*** through which the flow is determined by the ratio of ***throat*** pressure to atmospheric pressure.

BDC See ***bottom dead center***.

beach marks Fracture pattern indicative of progressive, long-term bending failure, so called because it resembles ripples formed by the sea on a sandy beach. Also called ***arrest marks***, ***conchoidal marks*** or ***striations***.

bead Part of a ***tire*** that seats onto a ***wheel rim***, normally strengthened with an embedded hoop or wire about which the casing cords are anchored. See also ***bead bundle***.

bead base Portion of a tire ***bead*** that seats on the ***bead seat***.

bead bundle Embedded hoop of wires or strong cords embedded in the bead of a tire, to prevent the tire lifting off the ***bead seat*** when inflated. Also ***bead coil***; ***bead cord***.

bead coil See ***bead bundle***.

bead cord See ***bead bundle***.

bead filler Solid rubber fillet that bonds the outer plies of a tire to the ***bead cord***. See Figures B.3 and R.2.

bead flange Fixed or removable lip on the outer periphery of a ***wheel*** that retains the tire. Also ***rim flange***. See also ***spring flange***. See Figure W.5.

bead heel Part of the bead of a ***tire*** that fits into the angle formed by the ***rim*** and ***rim flange***. See Figure R.2.

bead seat Part of a road wheel, below the ***bead lip***, on which the tire ***bead*** is seated.

bead toe Inner edge of ***bead base***.

beaded edge Any edge, as of a tire, upholstery, or body panel, in the form of a continuous lip or bead.

beam axle Rigid transverse beam on which nearside and offside wheels are mounted. The term originally described a ***dead axle*** of forged I-beam section but sometimes describes any rigid live axle or dead axle.

beam deflector switch See ***dip switch***.

beam indicator lamp Warning lamp on ***dash panel*** to indicate operation of headlamps. Also ***main warning lamp***; ***high beam warning lamp***.

bearing cap (1) Rigid, semi-circular retainer that locates and secures one half of a ***shell bearing***, as in an engine ***main bearing*** or ***connecting rod bearing***. See Figure B.1. (2) Removable disc or plug to prevent ingress of foreign matter into a bearing housing.

beaver back Rear body styling characteristic of some pre-war passenger cars in which a nominally vertical back is splayed into an outward sloping lower part.

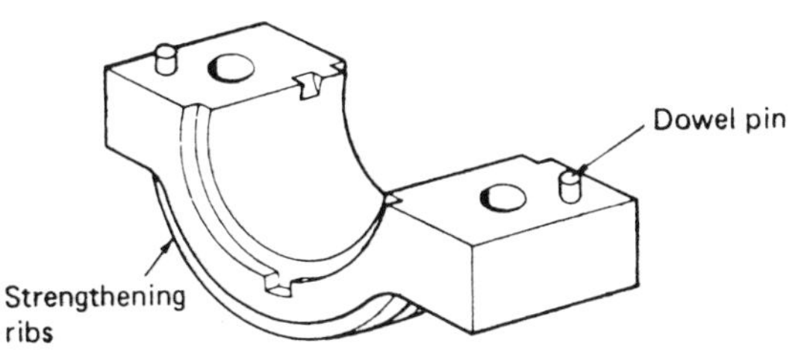

Figure B.1 Bearing cap for main bearing, showing dowel pins for accurate alignment with upper half of bearing in the crankcase.

beaver tail Ramped or lowering rear section of a ***low-bed trailer*** or ***low-loader***. Also ***tilt-bed trailer***. (Mainly US informal)

bellows See ***boot***.

Belgian block (UK: Belgian pavé) Road surface of convex-surfaced granite setts, used on proving grounds to assess vehicle integrity and durability, and response to road surface induced vibration.

Belgian pavé See ***Belgian block***.

bell housing Conical or bell-shaped extension of an engine crankcase, containing the ***flywheel*** and ***clutch***. See Figure H.2.

belt (1) Drive or power transmission belt such as a V-belt or toothed belt. (2) A driver or passenger safety belt. (3) Continuous reinforcing around the periphery of a ***radial ply tire***, usually woven from steel or a man-made fiber. Also ***breaker***. See also ***belt line***.

belt drive System of power transmission in which a flexible endless belt transmits power between pulleys. See ***fan belt***; ***van Doorn transmission***; ***toothed belt***; ***Variomatic transmission***.

belt idler See ***idler wheel***.

belt line The central or lower periphery of a car, particularly as a styling feature of sides and sometimes rear. Also ***waist line***.

belt slip Excessive slip of a power transmission belt such as a V-belt on a driving or driven pulley.

belted bias tire See ***bias belted tire***.

belted tire See ***bias belted tire***.

bench seat Wide vehicle seat for more than one person.

bench test Operating test carried out on an engine or other major item removed from the vehicle and mounted to facilitate observation and instrumentation. A static proving test.

Bendix drive A drive comprising a pinion wheel carried on a helically grooved shaft. The sudden rotation of the shaft causes axial movement of the pinion. Widely used to transmit ***starter motor*** power, the pinion engaging the ***ring gear*** teeth on the engine flywheel periphery. The sudden rotational acceleration caused by the firing of the engine throws the pinion out of engagement. See Figure B.2.

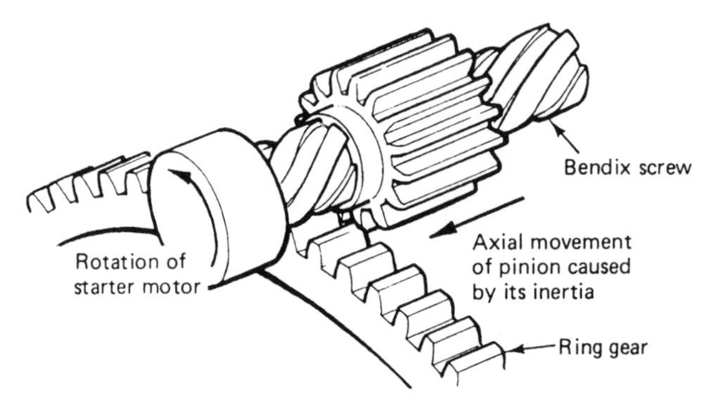

Figure B.2 Bendix drive of starter motor.

Bendix screw The coarse helix thread of the shaft of a ***Bendix drive***. See Figure B.2.

Bendix starter Starter motor employing a ***Bendix drive***.

Bendix-Tracta joint ***Constant velocity*** joint in which forked jaws are in sliding engagement with two slotted knuckles.

Bendix-Weiss joint A proprietary form of ***constant velocity joint*** in which the input and output elements are engaged by balls rolling in formed cavities. Both shafts require axial support. See also ***universal joint***.

benzene Organic chemical consisting or a hexagonal ring of carbon atoms with a hydrogen atom attached to each. Compounds containing one or more benzene rings are called ***aromatics***, and are an important constituent of ***petroleum***.

benzole Mainly aromatic hydrocarbon fuel ***additive*** and ***solvent***.

bevel differential Differential in which the principal differential gear elements are bevel gears.

bevel gear Gear wheel in which the teeth are cut onto a conical face rather than a parallel cylindrical face, so that two such wheels in mesh have shafts at an angle to each other, typically a right angle. See Figure D.5.

BHP See ***brake power***.

bi-focal headlamp Headlamp with a reflector with two focal lengths, normally used with single filament bulb and for dipped or lowered beam on four lamp systems.

bias belted tire Tire of ***bias ply*** or crossply construction but incorporating a substantially inextensible peripheral reinforcing ***belt*** such as used in a ***radial ply tire***. Also ***belted tire***; ***belted bias tire***. See also ***bias ply tire***.

bias ply tire (UK: crossply tire) Tire constructed on core or ***carcass*** of diagonally laid ***plies*** of fabric. See Figure B.3.

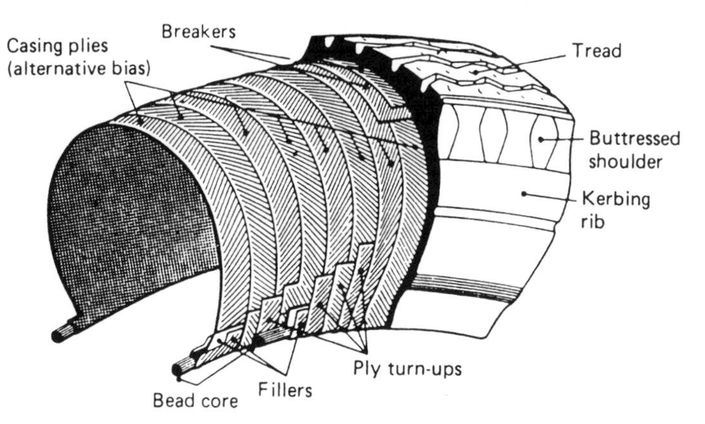

Figure B.3 Construction details of a bias ply or crossply tire.

biasing See ***variable torque dividing***.

big-end The end of a connecting rod that engages with a crankshaft. (UK informal) See also ***crankpin end***. See Figure C.10.

big-end bearing (US: connecting rod bearing) The ***connecting rod*** to ***crankshaft*** journal bearing. Before shell bearings were introduced, big-end bearings were cast on the connecting rod and reamed to size. On some types of engine, especially smaller two-stroke engines, rolling element bearings are used. Also ***rod bearing***. (US informal)

biodiesel Diesel fuel from biological or natural sources. See also ***rapeseed oil***.

Birfield universal joint Constant velocity joint that allows plunging action of one shaft relative to the other, named after manufacturer Birfield Transmissions Ltd.

blade connector Flat metal tongue-shaped electrical connector.

bleed (1) To empty a system of working fluid prior to maintenance or replenishment. (2) A valve or other means whereby a system can be drained of working fluid, or pressure reduced. (3) To remove entrapped air from a fluid system, leaving only the working fluid. See also ***purge***.

bleed screw Form of threaded tap or valve to facilitate draining of a hydraulic system, as for example a brake system.

blind spot Part of a vehicle's environment invisible to the driver in normal driving position, whether viewing directly or through a mirror.

block See ***cylinder block***.

blocker synchronizer Conventional transmission synchronizer with ***balk rings*** or blocking rings. (Archaic terminology)

blocking ring See ***balk ring***.

blow-back Sudden reversal of air flow through a ***carburetor***, often as a consequence of incorrect ignition or valve timing.

blow-by Unwanted leakage of gas under pressure, as past a piston or its sealing rings.

blow-out Sudden bursting of a tire.

blow-through turbocharger Turbocharger in which the compressor blows directly into the carburetor of a spark ignition engine.

blower (1) An exhaust ***turbocharger***. Originally a mechanically driven ***supercharger***. (Informal) (2) An interior ***cooling fan***. (Informal)

blown (1) Supercharged, turbocharged. (Informal) (2) Failed, usually catastrophically. (Slang)

bluff body Aerodynamic terminology for any body that is relatively wide in relation to its length, and particularly one with a blunt or squared front end.

boat tail Tapered rear termination of a vehicle, giving an appearance similar to that of the stern of a double-ended or canoe-sterned boat. (UK informal)

bob tail Articulated vehicle tractor operating without a trailer, particularly one with short rear overhang. (US informal)

body guides Fixed structural items that ensure accurate relocation of any removable load-carrying part of a commercial vehicle, such as a hopper, ***skip***, ***tipper body***, or ***demountable***.

body-in-white Stage in vehicle assembly comprising the assembled but unpainted panel-work, excluding trim and chassis items.

body-number (US: VIN) Manufacturer's bodywork identification number, normally unique to each vehicle.

body shell Assembled panelwork of a vehicle, and particularly a ***monocoque*** passenger car.

body tub Assembled panelwork of a vehicle, and particularly of an open or sports car, for mounting on a chassis.

bodywork The structural panelwork of a vehicle. See Figure B.4.

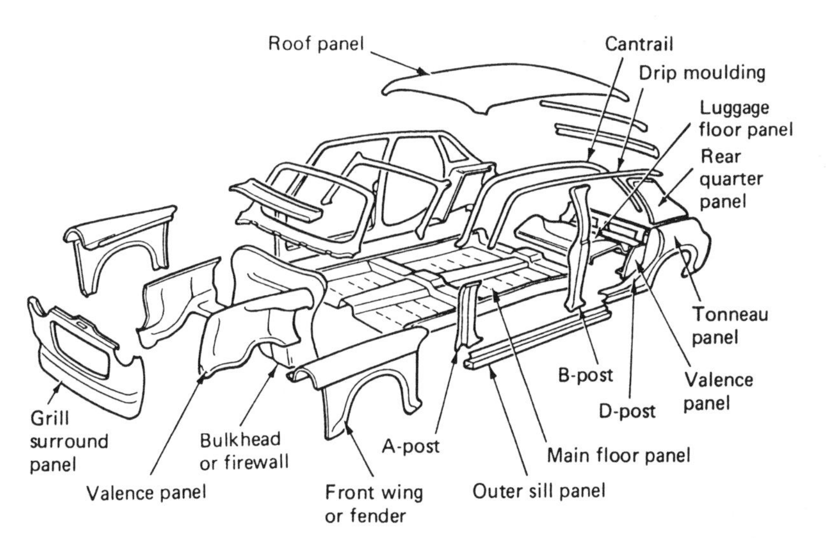

Figure B.4 Passenger car bodywork terminology (monocoque or unitary construction).

bogey Alternative spelling of ***bogie***.

bogie Heavy vehicle undercarriage in which ***tandem axles*** are mechanically linked and suspended as a unit. The term is sometimes applied to undercarriages of more than two axles, provided that the axles are linked by a compensating system. Also ***bogey***, particularly in US. See Figure T.1.

bogie spread Longitudinal distance between the outermost axle centerlines of a bogie undercarriage.

bolster Usually transverse horizontal commercial vehicle chassis member for carrying a distributed load, as for example a ***demountable body*** or load of poles.

bonnet (US: hood) Hinged body panel that gives access to the engine compartment of a vehicle, or to the forward luggage compartment if a rear-engined vehicle.

bonneted Of a commercial or public service vehicle, having the engine mounted forward of the cab and enclosed by a bonnet or hood. Also ***normal control*** (UK) or ***cab-behind-engine*** and ***conventional*** (US).

bonus loader Vehicle-mounted crane, forklift or other lifting device to facilitate loading.

boom (1) Acoustic effect of air flow over a vehicle, particularly where excited by apertures in the passenger compartment, such as open windows or sunshine roof. Also ***booming***. (2) Primary arm of earth-moving machine or crane. See Figure C.12.

booming See ***boom***.

boost (1) To increase, amplify or add to, particularly at a more vigorous rate than normal, as for example in boost charging a battery. (2) A measurable additional quantity, as of flow rate, pressure or electrical charge. (3) The additional pressure provided by a ***supercharger***, measured by a ***boost gage***.

boost gage (UK: boost gauge) Instrument for measuring ***supercharger*** output pressure.

boost gauge See ***boost gage***.

boost control Automatic control of fuel delivery, as in a turbocharged diesel engine when operating at lower engine speeds.

boost start Starting an engine of which the battery is discharged or feeble by applying a higher than normal voltage and current to the electrical system from an external source.

booster brake Brake used in addition to the ***service brakes*** of a vehicle to increase retardation under exceptional circumstances. See also ***brake booster***.

booster coil Auxiliary coil that increases the voltage or duration of the spark in a spark ignition engine.

boot (1) Thin, flexible elastomeric tube, often of concertina form, for protecting mechanical components, such as shaft couplings, from ingress of dirt and liquids. Also ***bellows***; ***gaiter***. (2) **(US: trunk)** Rear luggage compartment of a passenger car, internally isolated from the passenger compartment, and with a hinged lid for access. See also ***decklid***.

bore (1) The internal diameter of the cylinder of an engine or pump. (2) The cylinder wall of an engine. (Informal) (3) To make a cylindrical or circular aperture with a suitable cutting tool. (4) To increase displacement of an engine by increasing cylinder bore. (Informal) See also ***stroke***.

bore diameter See ***bore***.

bore-stroke ratio Ratio of bore to stroke. A ratio of 1:1 is referred to informally as square. See also ***oversquare***; ***undersquare***.

bottler's body (UK: brewer's dray) Normally enclosed goods vehicle for conveyance of crated bottles.

bottom board Floor panel of a van, particularly when removable. (UK) See also ***loadfloor***.

bottom yoke Structural item that links the forks of a motorcycle and reacts the loads of the bottom bearing of the ***steering head***.

bottoming (1) Deflection of a vehicle suspension to its maximum compressive travel due to road surface irregularity or excessive load. (2) Deflection of a pneumatic tire so that the carcass is momentarily flattened against the wheel rim. (3) Contact between the underside of a vehicle and raised ground.

bottom dead center The point of piston travel when the piston is nearest to the axis of the crankshaft. On a vertical engine, the lowest point of travel of the piston. Also ***BDC***; ***inner dead center***; ***lower dead center***; ***LDC***.

bottom end bearing The ***connecting rod*** to ***crankshaft journal bearing***. (US informal)

bottom gear The gear in a change-speed transmission that gives the lowest ratio of driven wheel rotational velocity to engine speed, and thus the highest torque at the driven wheels, as for starting, hill climbing. First gear.

bounce test Testing of the suspension, and particularly the ***shock absorbers***, of a vehicle by manually depressing and then releasing each corner and observing the decay of vibrations.

boundary lubrication Lubrication where the film thickness is of molecular depth and metal-to-metal contact is imminent. See also ***film strength***; ***hydrodynamic lubrication***; ***Stribeck Curve***.

Bowden brake Mechanical drum brake operated by a sheathed cable.

Bowden cable Form of mechanical control in which a multi-strand wire in tension operates within a flexible wire-wound outer sheathing, as frequently used for brake controls of motorcycles.

box Enclosed trailer or ***semi-trailer***. (US informal) See also ***horse-box***.

box-body Enclosed box-shaped freight-carrying body of a van or ***demountable***.

box section Closed section structural member capable of reacting loads in shear, bending and torsion.

box-van (US: panel body) Rigid, enclosed van in which a rectangular freight compartment is mounted behind the cab. See also ***luton***.

boxer engine A horizontally opposed engine. (Informal)

brake (1) Device to retard the motion of a vehicle or to prevent inadvertent motion when parked. (2) ***Dynamometer*** for measuring the power of an engine, originally a friction brake by which torque could be measured. Also ***brake dynamometer***. (Informal) (3) **(US: station wagon)** Passenger car body with rear doors and side lights, often with external wood framework, originally for carrying sporting equipment and dogs. A ***shooting brake***. More recently ***estate car***. (UK)

brake actuator See ***brake chamber***.

brake anti-roll Device to ensure that brake pressure is maintained when a vehicle is stopped on an incline. See also ***parking brake***.

brake booster Vacuum system for increasing brake pressure in a commercial vehicle. (Mainly US usage) See also ***brake servo***.

brake chamber Chamber containing diaphragm and ***pushrod*** to apply brakes in an air or ***hydraulic brake system***. Also ***brake actuator***, though strictly the actuator comprises the mechanical parts within the chamber. See also ***brake servo***.

brake drum Shallow closed cylinder attached to a vehicle ***wheel***, against the inner annular surface of which the braking effort is applied by a brake shoe faced with a friction material lining. See Figure C.1.

brake dynamometer Dynamometer in which power is absorbed by a mechanical or hydromechanical brake, by which torque can be measured.

brake fade See ***fade***.

brake fluid The fluid of a ***hydraulic brake*** system.

brake hop Hopping of wheels, particularly of a commercial vehicle or its trailer, when brakes are applied.

brake lamp See ***stop lamp***.

brake limiting valve Valve for limiting the pressure that can be applied to brakes, particularly of a commercial vehicle.

brake lining Friction material attached to the face of a ***brake shoe***, whereby the frictional force is applied by contact with the ***brake drum***. See Figure C.1.

brake mean effective pressure (1) Engine cylinder pressure, derived by calculation, that would give the measured brake horsepower. (2) The product of ***indicated mean effective pressure*** and mechanical efficiency.

brake pedal Foot-operated control by which the ***service brake*** is applied.

brake power Power developed by an engine as measured at the shaft by a brake or dynamometer. Specifically brake horsepower. See also ***indicated power***; ***pumping losses***.

brake retarder See ***retarder***.

brake servo Device for magnifying the ***brake pedal*** effort provided by the driver, usually by pneumatic or hydraulic means.

brake shoe The arcuate internally expanding element of a ***drum brake***, to which the friction lining is attached.

brake squeal High-pitched sound emitted by vehicle brakes.

brake valve (1) Valve operated by driver to pass air to air brakes at a pressure related to driver's pedal effort. (2) Foot-operated unit to provide graduated control of the brakes in a heavy vehicle service brake system.

braking ratio (1) Ratio of braking effort on front wheels to that on back wheels. (2) Deceleration of a vehicle as a proportion of the acceleration of gravity.

brass Alloy of copper and zinc, often with other elements and proportions of zinc up to 40 percent. Grades exist with good electrical and thermal conductivity, and moderate strength, and in cast and wrought forms.

break in (UK: run-in) To run new or reconditioned machinery under light load for a preliminary period.

breakdown truck See ***recovery vehicle***. A ***wrecker***. (Slang)

breaker See ***belt***.

breaker contact See ***contact breaker***.

breaker points See ***contact breaker***.

breakaway valve Valve which causes the brakes of a trailer to be automatically applied should the trailer become inadvertently separated from the towing vehicle. The valve will normally ensure that the brake system of the towing vehicle remains operative.

breakerless ignition See ***electronic ignition***; ***module***.

breakout box Terminal box for connection of electronic test instrumentation to relevant vehicle electrical and electronic circuits.

breather A vent to an enclosed container or case, as for example a ***fuel tank*** or engine ***crankcase***.

breeches pipe Y-configuration ***exhaust pipe*** forming confluence of two exhaust ***manifolds*** to one exhaust pipe, and resembling a pair of breeches, inverted. See Figure E.2.

bridge washer Washer that spreads load to its outer radius, particularly used for tire valve mounting on wheel rim.

bronze Alloy of copper and tin, often with other elements, of generally high strength, good thermal conductivity and corrosion resistance. See also ***phosphor bronze***.

Brookfield viscosity Viscosity or apparent viscosity measured by a Brookfield viscometer.

buck A model, often full size, of a vehicle for visual appraisal or as an aid to panel production. See also ***clay model***; ***studio buck***.

bucket seat Deep and rigid seat with side restraints to provide occupant constraint in fast cornering.

bucket tappet Cylindrical bucket-shaped ***tappet*** which encloses the end of a ***valve stem*** and retains the camshaft-end of the ***valve spring***. Sometimes refers to the lower ends of the pushrods when of such shape and function.

buddy seat (UK: sidecar) Single-wheeled carriage attached alongside a motorcycle.

buffer bar Horizontal bar mounted at front and rear of vehicle to prevent or reduce damage in low-speed impacts. Also ***bumper***.

built-up crankshaft Crankshaft assembled from separate component parts, for example with separate crankpins where a rolling-element connecting rod bearing is used.

bulker A large tipper or tanker, usually for carrying lower density bulk products. (Informal)

bulkhead (US: firewall) Transverse structural panel of a vehicle. See Figure B.4.

bull bars Sturdy, often tubular, framework for protecting lamps and other frontal items, particularly on all-terrain and leisure vehicles.

bulldozer Mobile earth-moving machine, often ***tracklaying***, equipped with a front-mounted controllable blade for moving and leveling earth. A "dozer." (Informal) See also ***angledozer***.

bullet connector Round-headed cylindrical electrical cable connector for insertion into an insulated connector tube. The male part of a ***snap connector***.

bullgrader A bulldozer equipped with a grading, usually angled, blade. See also ***angledozer***.

bump steer Change of ***steer angle*** resulting from sudden vertical deflection of the ***suspension***.

bump stop Compression spring, usually of rubber, that limits the deflection of a vehicle suspension on striking a bump. Also ***spring aid***.

bumper Horizontal bar mounted at front and rear of vehicle to prevent or reduce damage in low-speed impacts, and to absorb impact energy.

burner-heated catalyst Exhaust ***catalytic converter*** with facilities for preheating using a fuel-fed burner to reduce light-off time. Also ***BHC***. See also ***electrically heated catalyst***.

bush (US: bushing) Cylindrical sleeve forming a bearing surface for a shaft or pin.

butadiene Hydrocarbon which polymerises with itself to form rubbers. See also ***butyl rubber***.

butt A truck body. (UK regional)

butterfly carburetor Carburetor in which air-fuel flow is controlled by a disc throttle or butterfly. A ***fixed choke carburetor***.

butterfly valve Disc valve pivoted about its diameter and acting as a throttle in a pipe or chamber. A mild misnomer as the true butterfly valve consists of two hinged half-discs, opening and closing like the wings of a butterfly. See also ***throttle valve***.

butyl rubber ***Synthetic rubber*** produced by polymerising iso-butylene with butadiene or isoprene. Used as an alternative to natural rubber in ***tires*** and molded items. Noted for weather and heat resistance.

by-pass (1) Passage through which gas or liquid may flow instead of, or in addition to, its main channel, or any device for arranging this. (2) The transition system between the idle and main metering system in a ***carburetor***.

by-pass filter Filter which filters only a part of a circulating fluid at each pass.

by-pass valve A valve for directing flow through a by-pass.

C

C-matic transmission Trade name for semi-automatic clutchless transmission developed by Citröen.

cab (1) Compartment occupied by the driver, particularly of a commercial vehicle. (2) A taxi. (Informal)

cab air suspension System whereby the cab of a commercial vehicle is suspended pneumatically.

cab-alongside-engine Vehicle configuration in which the driver's cab is situated alongside the engine, notably in some off-highway vehicles and earlier designs of bus. Also ***CAE***.

cab-behind-engine (UK: normal control) Commercial vehicle in which the engine is mounted forward of the driver's cab. Also ***CBE***.

cab heater An auxiliary heater to heat the cab of a commercial vehicle, and particularly a **sleeper cab**, when the engine is not running.

cab-over-engine Vehicle configuration in which all or most of the engine is located beneath the driver's cab. See also ***forward control***.

cab-tilt See ***tilt-cab***.

cable brake Brake operated by cable and levers, either open cable in tension or ***Bowden cable***. Obsolete except for parking brakes.

cabriolet A ***soft-top*** or ***drop-head*** car.

cadmium Metal used mainly in protective plating of steel, but less often now because of health hazards.

cage (1) A retaining or locating piece for replicated mechanical items, such as ball or roller bearings. (2) A ***differential*** cage.

caliper In a disc brake system, the mechanism that brings brake pads to bear on disc by a clamping or pinching action, similar to that of the jaws of a vernier caliper. Sometimes ***calliper***. See also ***floating caliper***; ***swinging caliper***. See Figure D.6.

cam (1) A shaped or profiled component that produces linear or angular motion or ***lift*** of a ***follower*** while rotating. See Figure V.1. (2) The actuating cam of a drum brake, of various forms, and including an ***S-cam***. Except in the ***steering box***, many automotive cams are in the form of non-circular wheels which impart motion to the follower by their rotation. (3) A ***camshaft***. (Informal) Misleading usage, as for instance twin cam to describe an engine with two camshafts.

cam-actuated brake Type of ***drum brake*** in which the brake shoes are brought into contact with the drum by rotation of a cam. Also ***cam brake***. See also ***fixed cam brake***; ***floating cam brake***. See Figure C.1.

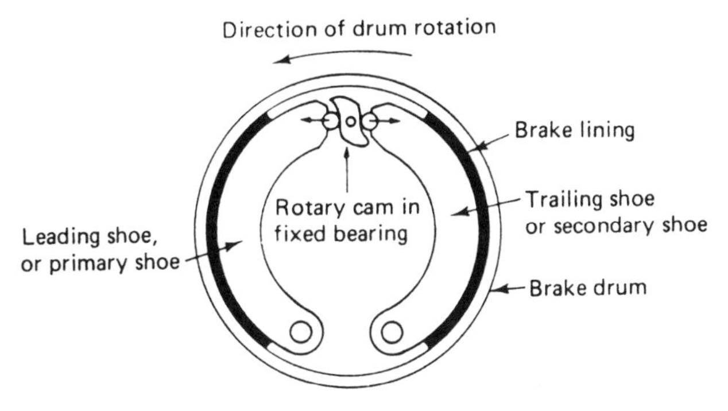

Figure C.1 A cam-actuated drum brake.

cam and lever See ***cam and peg***.

cam and peg Steering mechanism in which a conical peg mounted on a lever or rocker shaft engages in a helically cut cylindrical cam, the steering action being imparted by the lever. Also called ***cam and lever*** or ***worm and lever*** steering gear. (Obsolescent)

cam and roller steering gear Steering mechanism in which a tapered disc or discs (the rollers) engage with a helically cut waisted ***cam***, the steering action being imparted by the leverage about the steering drop arm shaft of the thrust on the faces of the rollers. See also ***Marles steering gear***.

cam brake Drum brake in which the shoes are actuated or spread by the rotation of a cam. Also ***cam-actuated brake***. See Figure C.1.

cam contour See ***cam profile***.

cam follower The part of a cam mechanism that rides on the contour surface of a cam. See ***tappet***; ***valve lifter***.

cam ground piston A piston ground to a slightly out-of-circular section to counteract thermal distortion. Also ***oval piston***. (Informal)

cam profile The shape of the periphery of a ***cam***. The contour determines the stroke (throw) and linear acceleration of the follower.

cam roller Cam follower in the form of a rotating wheel.

camber (1) Convex arched curvature of a (usually horizontal) surface. (2) Average curvature of the chordwise section of an aerofoil. (3) Mildly arched profile of a road or pavement. (4) Inclination of the plane of a wheel to the vertical plane of symmetry of a vehicle. Camber is considered positive if the wheels lean out towards the top, and negative if they slope inward. See Figure C.2. See also ***camber angle***. (5) Curvature of a leaf spring.

camber angle Angle between the plane of a wheel and the vertical axis of a vehicle when viewed in end elevation. Normally quoted under specified load or ride height and with steering ahead. ***Rake angle***. (Informal) See Figure C.2.

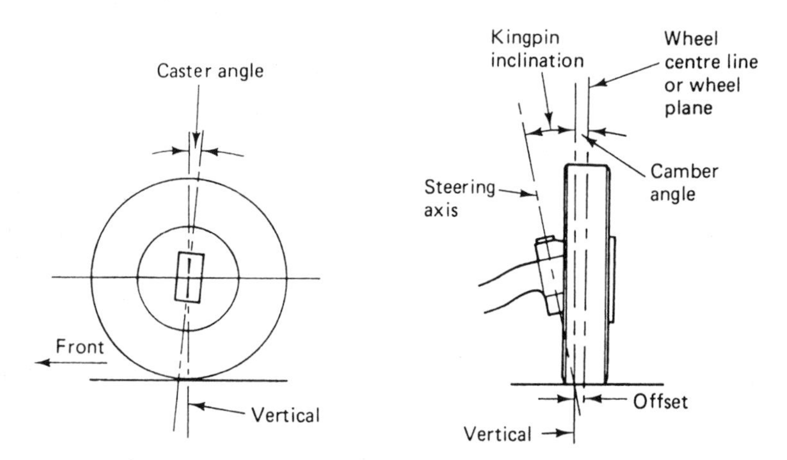

Figure C.2 Left: side view of steered wheel. Right: steered wheel viewed from forward; the illustration shows a wheel with a small positive camber.

camber stiffness Rate of change of lateral force on ***tire*** with change in ***camber angle***.

camber wear Wear pattern of ***tire*** in which ***tread*** on one side of the tire is evenly worn.

camper Touring vehicle with van-type body with sleeping and residential facilities. A ***motor caravan***.

camping trailer Low-profile trailer unit from which temporary recreational accommodation can be erected at a campsite.

camshaft Shaft on which suitably phased ***cams*** are mounted, as for example to operate intake and exhaust valves of an engine. See Figures E.1. and V.1.

camshaft pump (1) A pump, as for oil or cooling water, driven directly from a cam on the camshaft of an engine. (2) A reciprocating pump in which the piston is moved by a cam rather than by a connecting rod and crank.

cancellation In acoustics, the reduction of noise by the application of tuned resonators, as for example to an ***induction system***. See also ***absorption***.

cantilever spring Suspension spring, normally a ***half-elliptical leaf***, rigidly attached to the vehicle at its major section and carrying the undercarriage at its end section.

cantrail Upper longitudinal structural member of a vehicle body, and particularly of a box-van, to which the ***roof panel*** and ***roofsticks*** are attached. Also ***cant rail***. See Figures B.4 and C.13.

capacitor Electrical component which stores an electric charge when subjected to a voltage difference. Also ***condenser***, although this term is considered archaic in electrical and electronic circles, but still widely used in automotive context.

capacitor discharge system Ignition system which stores its primary energy in a capacitor. Also ***capacity discharge system*** or ***CD system***.

capacity (1) In an engine, the product of the ***bore*** cross-sectional area, the piston ***stroke*** and the number of cylinders. Expressed in Europe as cubic centimetres or litres, and in the US as cubic inches. See also ***displacement***; ***swept volume***. (2) The volume or usable volume of any reservoir or container, such as a ***fuel tank***. (3) The capacitance of an electrical capacitor or condenser. (Obsolete)

capsize mode Directional perturbation, often oscillatory, that if unstable may lead to the overturning of an uncontrolled vehicle.

car (1) A private motor passenger vehicle. A motor car or automobile. (2) A single vehicle running in tracks, as for example a carriage of a train, or a tramcar. (3) A passenger ***coach***. (Mainly local usage, from French)

caravan Towed vehicle, usually with one axle at or near balance point, equipped for accommodation. (Mainly UK usage)

carbon (1) Chemical element, which can exist on its own in several forms, such as amorphous carbon, graphite, diamond and coal. It is present in all organic compounds, and when combined chemically with hydrogen, forms ***hydrocarbons***. (2) Principal (but not only) element of deposit found on piston crown and ***cylinder head*** of engine, resulting from incomplete combustion. Informally referred to as ***coke***.

carbon monoxide Gaseous product of incomplete combustion of ***hydrocarbon*** fuel with air. One of the principal toxic pollutants of the exhaust, particularly of spark ignition engines. Chemical formula CO. See also ***emission***; ***emission control***.

carburetion (UK: carburation) The vaporising of a fuel and mixing of the vapor in appropriate proportions with a stream of air in a carburetor. See also ***stoichiometric***.

carburetor (UK: carburettor, carburetter) Device for vaporising liquid fuel and mixing it in appropriate proportions with a stream of air prior to combustion in an engine. See Figure C.3.

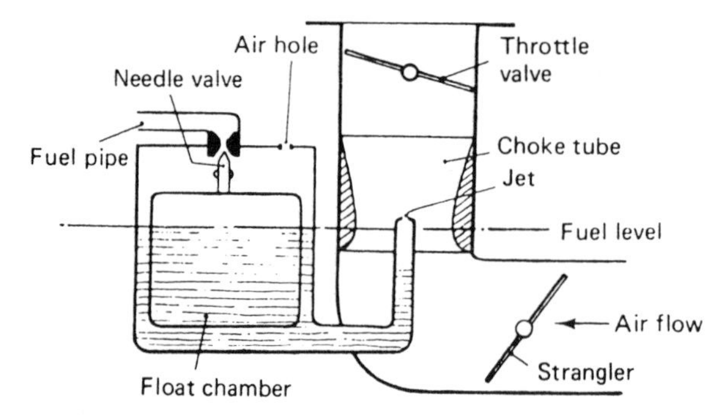

Figure C.3 A simple updraft carburetor.

carcass Structural body of a ***tire*** built to resist air pressure and to which the rubber ***tread*** and ***sidewall*** rubber are bonded. The term ***carcass*** may sometimes exclude the ***bead***. Also ***casing***. See Figure B.3.

carcass ply Tire ply extending from ***bead*** to bead. See also ***stabilizer ply***.

carcinogen Substance which is known to cause cancer.

cardan axle Axle with one or more cardan joint, as for example the drive shafts of a ***de Dion*** system. Sometimes Cardan axle. See Figure H.2.

cardan click Clicking sound produced by double Cardan or Hooke's joints, either in forward or reverse drive, generally at low speed, and on ***steering axles*** often exacerbated by steering input.

cardan drive See ***cardan shaft***.

cardan joint Universal joint in which a cruciform member couples two yokes. Also ***Hooke's joint***. See Figure H.1.

cardan shaft Propeller shaft fitted with universal joints at each end. Also ***cardan drive***. See Figure H.2.

cardan tube A tubular ***cardan shaft***.

cargo floor Commercial vehicle or trailer floor surfaced and stressed for carrying cargo. Also ***load floor***.

carrozeria house A studio or design office specializing in the design of vehicle bodywork and styling.

casing (1) Reinforcing structure of a ***tire***, to which the extendible rubber ***tread*** and ***sidewalls*** are attached. See also ***carcass***. (2) Protective cover, as of an instrument.

caster (1) Originally an undercarriage member with a vertical pivot axis behind which the wheel trailed, so that the wheel aligned itself to its direction of motion. (2) Longitudinal distance at ground between projected vertical plane through wheel

spin axis and projected point of intersection of ***kingpin axis*** with ground, giving the effect of caster as described in 1. Also ***caster offset*** and ***caster trail***. Castor (alternative spelling). See also ***caster angle***. See Figure C.2.

caster action Self-centering steering action, particularly when attributable to caster.

caster angle Angle in side elevation between the steering and vertical axes. See also ***self-aligning torque***. See Figure C.2.

caster offset Longitudinal distance between intersection of ***steering axis*** with the ground and the ***center of tire contact*** (or intersection with ground of vertical through wheel ***spin axis***). Offset is considered positive when the intersection point is forward of the ***center of tire contact***. Also ***mechanical trail***.

cat ***Catalytic converter***. (Informal)

catalytic converter Emission control device fitted in the ***exhaust system*** of an internal combustion engine. The converter reduces the toxicity of products of combustion by catalytic re-combination. See ***emission control***.

Caterpillar track Proprietary linked metal chain of track laying vehicle, after name of manufacturer. Term often loosely used to describe any form of ***track laying*** mechanism. See Figure C.12.

cell (1) Any single unit, either singly or as part of a ***battery***, for storing or releasing electricity by electrochemical reactions. (2) The combustion chamber of a ***rotary engine***, particularly where non-cylindrical, as in a ***Wankel engine***.

cell stack A package of connected electrical cells, not necessarily in one integral casing.

cell swept volume Difference between maximum and minimum cell volume of a ***rotary engine*** as it rotates. Variously of one cell or the total of the three cells attending each rotor of a ***Wankel engine***. See also ***geometric displacement***.

cement mixer Vehicle for the transportation of wet cement, and able to turn the cement while in transit.

center-axle trailer Towed vehicle in which the axle or axles are located near to the center of gravity, the trailer depending on the towing vehicle for fore-and-aft (pitch) stability. See also ***caravan***.

center of parallel wheel motion Center of curvature of path along which each pair of wheel centers moves in a longitudinal vertical plane relative to the sprung mass when both wheels are equally displaced. See also ***instantaneous suspension center***.

center of tire contact Intersection of ***wheel plane*** and vertical projection of the ***spin axis*** of the wheel onto the road plane.

center plate The driven plate of a ***clutch***, which is sandwiched between the ***flywheel*** and the ***pressure plate***. See Figure C.4.

center point steering Steering geometry in which ***steering axis*** meets plane of steered wheel at ground level, that is, with no ***lateral offset***. For most practical purposes this implies that the steering axis passes through the longitudinal centerline of the tire contact patch. See also ***kingpin offset***; ***negative offset steering***; ***positive offset steering***. See Figure C.2.

center wear Type of irregular ***tire*** wear characterised by increasing wear from ***shoulder*** to center of tread band. See also ***shoulder wear***.

central chassis lubrication Grease or oil lubrication of chassis components (excluding engine, gearbox, etc.) from one central pressurized reservoir.

central pillar Nominally vertical structural support of the roof, against which the front door closes. Also ***B-post*** or ***B-pillar***. See Figure B.4.

centrifugal advance A mechanical device that employs rotating masses in a similar way to a mechanical governor, to advance the spark timing in a spark ignition engine.

centrifugal caster (1) Self-centering effect brought about by rotation of a wheel. (2) Unbalance moment about the ***steering axis*** produced by lateral acceleration equal to gravity acting at the combined center of gravity of all the steerable parts. Considered positive if the combined center of gravity is forward of the steering axis and negative if rearward. (SAE definition)

centrifugal clutch Friction clutch in which friction elements are brought into contact with the clutch face by "centrifugal force." The clutch will therefore automatically engage and disengage at a certain speed of rotation. A centrifugal clutch may be of a drum (direct acting) type or disc type, in which weights actuate a lever mechanism.

centrifugal filter Filter in which particulate matter is removed from a flow of liquid, such as oil or fuel, by the action of centrifugal force brought about by rapidly rotating the liquid flow in a spinning element.

centrifugal ignition advance See ***centrifugal advance***.

centrifugal turbocharger Turbocharger in which the turbine and compressor are of "centrifugal" or radial flow design as opposed to axial. Most turbochargers are of this type.

cetane Paraffinic hydrocarbon and primary reference fuel on which the cetane number scale for measuring the ignition quality of diesel fuels is based. It has a ***Cetane Number*** of 100.

cetane improver Diesel fuel ***additive*** for improving the ***Cetane Number***.

Cetane Index Rating for diesel fuels calculated from an empirical formula with API gravity and volatility parameter, rather than from tests on a standard engine. The method can yield misleading results as it may not acknowledge the effect of ***cetane improvers***.

Cetane Number Measure of the ignition quality of a diesel fuel, determined from tests on a standard engine. It is a measure of time required for a fuel to ignite after injection. A high ***Cetane Number*** indicates a short lag. See also ***Cetane Index***; ***Diesel Index***.

CFR engine Variable compression test engine developed under the direction of the Cooperative Fuel Research Committee (CFR) and by which ***Antiknock Index*** of fuels can be determined.

chafer Protective surface of tire ***bead*** that prevents damage to the bead on changing tires. Also ***clinch strip***; ***rim strip***.

chain (1) A series of interlinked ovoid metal rings forming a length or continuous loop. (2) A series of pairs of links interconnected at their joints by parallel rods which may carry, in the case of a ***roller chain***, hardened steel rollers. (3) A roller chain consisting of two or more parallel sets of links and rollers, as in the ***duplex chain*** or ***triplex chain***.

chain and sprocket drive Power transmission in which a ***roller chain*** engages with two or more toothed wheels or sprockets. Used mainly on bicycles, motorcycles, and in engines as a drive from crankshaft to camshaft.

chain case Case to enclose a ***chain and sprocket drive***.

chain drive See ***chain and sprocket drive***; ***roller chain***.

chain driven A vehicle or other mechanism in which power is transmitted by a ***chain and sprocket drive***.

chain guard Partial housing, casing or cover to shield a chain and protect the user from soiling by oil or grease.

chain tensioner Mechanically or hydraulically operated device to maintain tension in a chain.

chain track Metal linked track of a ***track-laying vehicle***. See also ***Caterpillar track***.

chain wheel Any wheel around which a chain runs, but particularly a toothed sprocket wheel.

chalking Surface deterioration of paints, certain ***elastomers*** and ***polymers***, typically due to the effect of the ultraviolet content of sunlight or oxidation.

change down (US: downshift) To select a lower gear.

change-speed gearbox (US: transmission) Set of movable or ***constant mesh*** gears permitting the speed ratio between input and output shafts to be changed either manually or automatically. See Figure G.1.

Chapman strut Suspension system comprising a telescopic strut with its uppermost end attached to the chassis, and the lower end constrained in the lateral and longitudinal planes by two links, as for instance a ***wishbone***. After Colin Chapman, sports and racing car designer. A development of the ***MacPherson strut***. Generally used for rear suspensions. Term falling into disuse.

charge air cooling Cooling of the air charge downstream of a ***turbocharger*** or ***supercharger***.

charabanc A motor coach. (Archaic, from French *char-à-banc*, coach with bench seats)

charcoal canister Trap containing charcoal granules to store fuel evaporating from a fuel system and prevent its loss to atmosphere, particularly from a ***carburetor*** and fuel tank.

charge (1) Usable electrical capacity, as of a storage ***battery*** or ***capacitor*** (condenser). (2) To increase the electrical charge in a battery. To charge a battery, using a battery charger, dynamo or alternator. (3) The quantity or mass of fuel (or air and fuel) entering the cylinder of an engine on each ***stroke***.

charge cooling Removal of heat from the induction charge of an engine to increase its density and consequently the total charge mass and engine output per firing stroke. See also ***intercooler***.

charge current (UK: charging current) Electrical current flowing to the ***battery*** when charging.

charging stroke The stroke of an engine in which the charge enters the combustion chamber. The ***induction*** or inlet stroke.

chassis Structural lower part of a vehicle to which the ***running gear*** and body is attached. The true chassis is now evident only in heavy goods vehicles, some public service vehicles, and some specialist cars.

chassis cab Commercial vehicle chassis complete with cab, engine and ***running gear***, and equipped for road use, but without body or load platform. See also ***chassis cowl***.

chassis cowl Commercial vehicle chassis complete with engine and ***running gear***, and equipped for road use, but with only the front of the cab, from which a body builder can construct an integral van such as a ***pantechnicon*** or ***horse-box***. See also ***chassis scuttle***.

chassis dynamometer Test equipment fitted with rollers for the wheels of one or both axles of a complete vehicle, capable of providing drive input and measuring output parameters such as power and torque at the wheels.

chassis scuttle ***Chassis cowl***, without ***windshield (windscreen)***.

chassis stop Bracket or other robust protrusion from the chassis, particularly of a heavy vehicle, against which a leaf ***helper spring*** bears when the main spring is greatly deflected.

chatter Any irregular or jerky motion of a component, as for example a ***wiper blade*** over glass, or the noise associated therewith.

check valve Hydraulic valve that maintains a residual pressure, as in a brake system when the brake is not applied. A ***load-holding valve*** that permits flow in one direction only. See also ***shuttle valve***.

checking Pattern of small surface cracks, usually on a rubber surface exposed to weather or sunlight. (Mainly US usage) Also ***crazing*** (UK).

chilled distribution Carriage of goods at a low, controlled temperature, usually between 0 and 5 degrees Celsius. See also ***reefer***; ***refrigerated vehicle***.

Chinese six Six-wheeled rigid truck with two front ***steering axles*** and one rear ***driven axle***. (Informal)

chlorofluorocarbon Chemical, normally in liquid or vapor form, used as a secondary refrigerant in refrigeration and air conditioning systems. Also ***Freon***.

choke Valve that restricts the amount of air entering an engine on the ***induction stroke***, thereby enriching the fuel:air ratio for ease of starting and running when cold. ***Strangler***. (Informal) See also ***automatic choke***.

choke stove Heat exchange chamber on ***exhaust manifold*** to speed the operation of an ***automatic choke***. (Mainly US usage)

choke tube A ***carburetor*** venturi. (US informal) See Figure C.3.

choke valve Valve that restricts the flow of a gas or liquid.

choked Of an engine, running with the choke in operation.

chrome Popular, but unscientific, name for ***chromium***.

chromium Silvery white, relatively soft metal with high resistance to atmospheric surface degradation. Used as a decorative plating on mainly steel components, though processes exist facilitating plating onto plastics. Chemical symbol Cr.

chronic toxicity Toxic effect resulting from long exposure to a mildly toxic material.

chunking Loss of pieces of tire ***tread*** while a vehicle is in motion.

CI engine A compression ignition engine. See also ***diesel engine***.

cill See ***sill***.

circuit breaker Device for interrupting an electrical circuit when the current exceeds a predetermined value. The device can be reset after the circuit malfunction has been corrected.

city cycle Any standard vehicle test cycle that simulates urban driving, with frequent gear shifts, and driving predominantly in the lower gears. See also ***highway cycle***; ***urban cycle***.

clay buck See ***clay model***.

clay model Full size model made prior to manufacture, by which styling, panel molding and other qualities are assessed. Also ***clay buck***. See also ***studio buck***.

clearance lamps Lamps mounted on the permanent structure of a large vehicle near to the upper left and right extreme edges to indicate the overall width and height of the vehicle. See also ***marker lamp***.

clearance volume Volume remaining above the piston of an engine when it reaches ***top dead center***. See also ***swept volume***.

clerestory head Engine cylinder in which often axially opposed ***intake valves*** and ***exhaust valves*** open into an extended chamber of smaller cross section than the cylinder bore.

clicker wheel adjuster Adjusting mechanism with toothed indexing wheels, used for example for ***brake*** adjustment.

climatic tunnel See ***climatic wind tunnel***.

climatic wind tunnel Wind tunnel with facilities for simulating climatic influences such as temperature and humidity, and for assessing heating, cooling and air conditioning systems. Also ***climatic tunnel***.

climatic wind chamber Test facility that simulates air flow and climatic conditions. See also ***climatic tunnel***.

clinch Sturdy clip that holds together the leaves of a ***leaf spring***, usually riveted to one leaf thus permitting relative movement of other leaves. See also ***leaf spring strap***.

clinch strip See ***chafer***.

clocking The (illegal) tampering with the ***odometer*** reading of a vehicle. (Informal)

close-coupled city bus An articulated single-deck bus in which tractor and trailer (or semi-trailer) carry passengers, and may be coupled so that there is interconnection between tractor and trailer.

close-coupled trailer A two-axle trailer, with both axles mounted close together, nominally within one meter of each other, neither axle having facility for steering.

closed cycle turbine Gas turbine in which the working gas is recirculated rather than exhausted to atmosphere.

closed-loop dwell-angle control Electronic feedback system that maintains ***ignition*** primary current under varying conditions of battery charge.

closed-loop engine control system Automatic control of engine parameters through direct feedback of data, as from electronic sensors. See also ***limit cycle control***; ***proportional control***.

closed-loop fuel system Electronically controlled ***carburetor*** or ***fuel injection*** system in which mixture strength is adjusted by feedback signal from the monitoring of exhaust gas composition, as for example the oxygen or unburned hydrocarbon content of the exhaust.

Cloud Point Temperature at which ***wax*** is no longer held in solution in an oil or fuel as it is cooled, resulting in a cloudy appearance. See also ***wax plugging***.

cloverleaf head Engine cylinder head in which the disposition of the valves, with a central ***spark plug***, gives the appearance of the leaf of a clover.

cluster gear (1) Set of gears fixed to a ***layshaft*** (UK) or ***countershaft*** (US). (2) Layshaft or countershaft gear assembly. See Figure G.1.

clutch Mechanism for engaging or disengaging the transmission of power between two axial shafts. In vehicle manual transmission systems the clutch allows smooth and progressive engagement between engine and ***gearbox*** for taking up initial drive, and rapid disengagement for gear selection. Clutch terminology is far from standardized, varying considerably from manufacturer to manufacturer. See also ***centrifugal clutch***; ***cone clutch***; ***dog clutch***; ***fluid flywheel***; ***freewheel***; ***multi-plate clutch***. See Figure C.4.

clutch brake Brake that slows the rotation of a disengaged clutch ***driven plate*** to make gear engagement easier, particularly on large capacity clutches. Also ***clutch stop***.

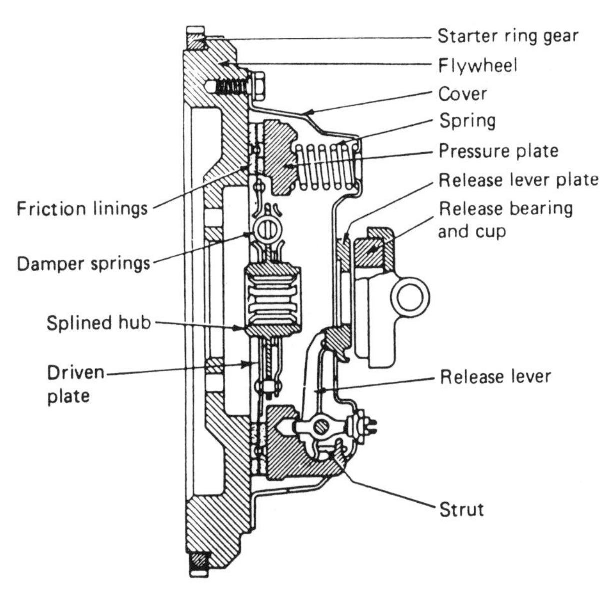

Figure C.4 Push-type clutch mechanism.

clutch cover Bowl-shaped cover that houses the rotating elements of a friction clutch, anchors the clutch springs, and is normally attached to the flywheel, rotating at engine speed. See Figure C.4.

clutch disc (UK: clutch plate, clutch center plate or **driven plate)** Rotating clutch element, to which the friction material is attached. See also ***cushion drive***.

clutch drag Incomplete disengagement of a clutch, so that some drive torque is transmitted when the ***clutch pedal*** is fully depressed, making gear changing difficult.

clutch driven plate (US: clutch disc or **disk)** Rotating clutch element, to which the friction material is attached. See also ***cushion drive***.

clutch fluid Oil used in the actuating system of a hydraulically operated clutch.

clutch housing The outer casing of the clutch assembly. The housing often forms a structural link between engine ***crankcase*** and ***gearbox***. See also ***bell housing***.

clutch judder Faulty condition in which the clutch does not take up smoothly but vibrates during engagement due to oil on friction surfaces or to warped plates. Also ***clutch shudder***.

clutch pedal Pedal by which the driver of a vehicle operates the clutch, through a mechanical or hydraulic linkage. See Figure C.5.

clutch pedal free travel Distance moved by clutch pedal before clutch actuating mechanism begins to operate.

clutch plate (US: clutch disc or disk) Rotating clutch element, to which the friction material is attached. See also ***cushion drive***; ***solid drive***.

clutch pop Sudden engagement of clutch, particularly to an engine already throttled to develop power. (US informal)

clutch pressure plate Robust metal disc which holds the friction-lined ***clutch disc*** (plate) by spring pressure against the engine flywheel and thereby causes engine torque to be transmitted to the transmission system under spring pressure. Sometimes incorporated in the clutch cover. Also ***pressure plate***. See Figure C.4.

clutch release bearing (US: throwout bearing) Shaft-mounted thrust bearing that transmits the action of the clutch pedal to disengage a clutch. See also ***clutch release lever***. See Figures C.4. and C.5.

clutch release lever Lever, mechanically or hydraulically actuated by clutch pedal, that acts on clutch release bearing to disengage clutch. Also ***release fork***; ***throwout fork***; ***thrust bearing actuating lever***. See Figures C.4. and C.5.

clutch release mechanism The mechanical and hydraulic parts that release the pressure between ***driven plate*** and ***pressure plate*** to disengage the clutch. See also ***throwout bearing***. See Figure C.5.

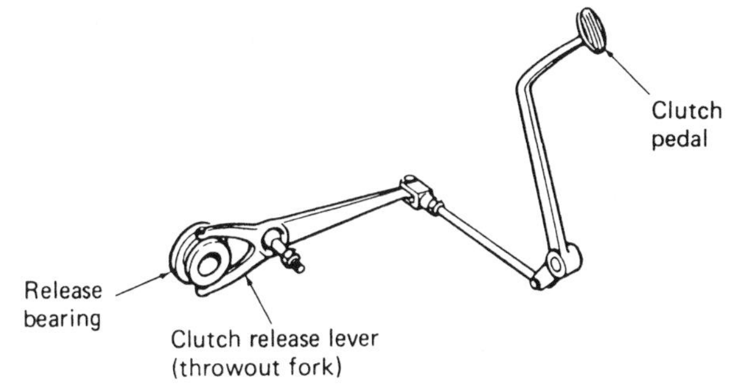

Figure C.5 Mechanical clutch release mechanism.

clutch servo Compressed air-operated servo mechanism that assists the hydraulic operation of a clutch ***slave cylinder*** on heavy vehicles.

clutch shaft Shaft through which power is transmitted from ***clutch driven plate*** to ***gearbox*** or transmission input shaft. Also drive pinion. (US informal)

clutch shudder See ***clutch judder***.

clutch slip Faulty condition in which the friction in the clutch is insufficient to prevent relative movement of driving and driven plates under power, so that engine speed rises without a corresponding increase in road speed.

clutch springs Coil springs (or diaphragm spring) that provide the clamping force between pressure plate and driven plate in a clutch. Also ***pressure plate springs***; ***thrust springs***.

clutch start (1) To start an engine of a manual transmission vehicle by pushing or towing the vehicle with the clutch disengaged and a high gear engaged, then engaging the clutch so that the engine is turned by way of the transmission system. (2) In motorcycle racing, a standing start with engine running rather than a run-and-bump start.

clutch stop A brake, operated by the clutch pedal, to slow rotational speed of a manual transmission (gearbox) primary shaft and so facilitate upward shifting of gears, particularly in commercial vehicles with ***crash gearboxes***. (Obsolete)

coach A usually single-deck multi-passenger vehicle, mainly for touring or long-distance operation and not operating between stages, as would an omnibus. Mainly British usage. The term ***car*** is occasionally used in the UK.

coachwork The panelled bodywork of a vehicle. See Figure B.4.

coarse acceleration Heavy use of the accelerator pedal, generally carried out for effect, and likely to induce ***wheelspin***.

coarse braking Unnecessarily sudden and aggressive braking, likely to induce ***skidding*** or wheel locking.

coarse steering Unnecessarily sudden and aggressive steering, likely to induce loss of adhesion under conditions of reduced road friction.

coast-down (1) Test procedure for deriving total resistance from deceleration, from which, for example, aerodynamic drag can be deduced from data on rolling resistance. (2) Deceleration test for gaseous emissions or noise level. Also ***coastdown test***. See also ***drawbar test***.

cocktail effect Environmental effect where a mix of pollutants may have a more injurious effect on health than the individual pollutants in isolation.

cocktail shaker piston Piston incorporating an oil chamber beneath the ***undercrown***. Contact between the oil and undercrown provides additional cooling to the ***piston crown***. See Figure C.6.

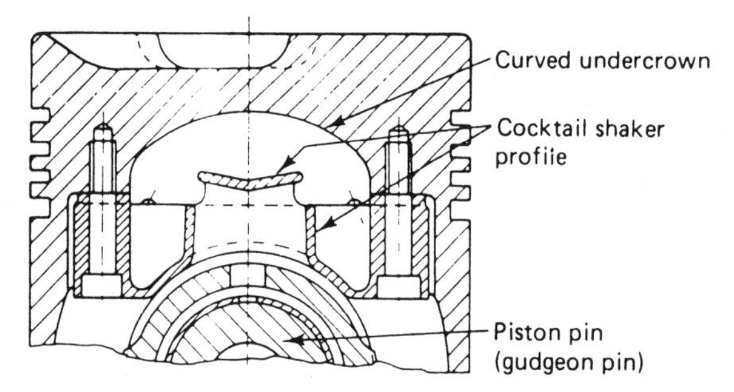

Figure C.6 Upper part of cocktail shaker piston; oil supply to the shaker is replenished while the engine is operating.

cocoa Metallic oxides generated by fretting, and similar in appearance to rust. (Informal)

cogged belt A toothed rubber ***timing belt***. (Informal)

coil See ***ignition coil***.

coil and wishbone (1) Independent suspension in which the wheel assembly is located by wishbones, with a coil spring as the springing medium. (2) Suspension comprising a ***beam axle*** suspended by two coil springs and constrained longitudinally and transversely at each side by ***wishbones***.

coil-over ***Coil spring*** coaxial with a suspension damper. See Figure P.5.

coil spring A helical spring, mainly used in vehicles for reacting compressive loads, as in ***suspensions*** or ***poppet valve*** mechanisms.

coil spring clutch Clutch in which the ***friction plate*** is held in position by a set of coil springs acting around its outer edge. See Figure C.4.

coke Carbon deposit in engine. (Informal)

cold cranking rating Current drain of battery when ***starter motor*** is turning a cold engine for a specified time.

cold plug Spark plug that operates at a low temperature in relation to the combustion temperature, having a relatively high heat conductivity. Also ***hard plug***.

cold press molding System for making GRP panels in matching pairs of tools which are brought together under mechanical pressure or under the weight of the (usually concrete) tools. The process is suitable for production quantities of two thousand or more. See also ***hand lay-up***; ***hot press molding***.

cold start ballast resistor See ***ballast resistor***.

cold start device Any device, such as a ***choke*** or additional fuel injector, that temporarily enriches the flow of fuel to an engine to facilitate starting from cold.

cold start injector See ***cold start device***.

cold sticking Sticking of ***piston ring*** in groove when engine is cold. The ring will normally become free when the engine is warm. See also ***hot sticking***.

cold testing Testing of an engine without the engine running under its own power, for example to test integrity of mechanical and lubrication systems.

colloid Suspension of microscopically fine particles in a liquid which do not settle out and are not easily filtered out.

column shift (UK: column change) Gearshift lever mounted on the ***steering column***.

colza oil See ***rape oil***.

combination Vehicle consisting or two or more separable units, of which each part need not be independently mobile, for example a motorcycle and ***side-car*** (buddy seat) combination or a commercial vehicle ***tractor*** and ***semi-trailer*** combination. See Figure A.4.

Combined Fuel Economy Standard measure of fuel economy derived from a stipulated proportion of urban and highway cycle driving.

combined retarder An integrated ***retarder*** equipped with a manually operated deselect device.

combustion chamber The part of an engine in which combustion takes place, normally the volume of the cylinder between ***piston crown*** and ***cylinder head*** in a reciprocating engine. In ***indirect injection*** compression ignition engines and in certain designs of spark ignition engine in which combustion is initiated in a separate cell or cavity, the cell or cavity is not necessarily considered to be part of the combustion chamber.

combustion chamber surface:volume ratio Ratio of surface area of the ***combustion chamber*** to its volume at a stated ***stroke*** of the piston. The ratio influences the ***quenching*** effect of the cylinder walls and the local formation of ***unburned hydrocarbons***.

combustor The combustion chamber of a gas turbine. The chamber between compressor and turbine into which fuel is sprayed and burned. Also ***combustion chamber***.

Comet head Indirect injection combustion system for diesel engines, developed by Ricardo. See also ***prechamber***. See Figure C.7.

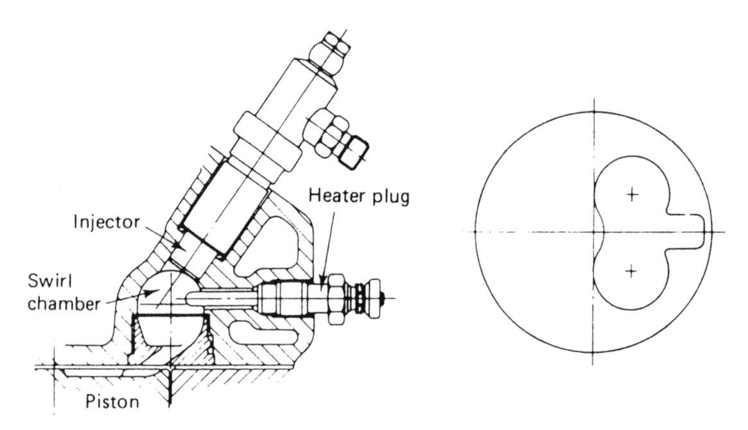

Figure C.7 The Ricardo Comet swirl chamber and head.

commercial oils Oils for use in commercial vehicle diesel engines, to specified national or international categories.

commercial vehicle Vehicle used for the transportation of goods, materials, or for gain or reward other than in the conveyance of passengers, and registered or licensed as such. Exact definition will be subject to local variation and interpretation. Also ***CV***. See also ***articulated vehicle***; ***goods vehicle***; ***lorry***; ***public service vehicle***; ***truck***; ***van***.

common rail Fuel injection system in which fuel in supplied to each ***injector*** from one high-pressure source by way of a linking passage or "common rail." Mainly used for stationary and marine engines. See also ***jerk pump***.

compensating axle suspension ***Tandem axle*** suspension in which there is a positive transfer of load between axles, for example by mechanical, pneumatic or hydraulic interconnection. Also ***reactive suspension***. See also ***balance beam***; ***walking beam***. See Figure T.1.

compensating jet (US: air bleed passage) Jet or passage in a ***carburetor*** with a branch to atmosphere through which air is drawn and introduced into the fuel flow at high flow rates, thus preventing over-richness. See Figure C.8.

compensating linkage See ***compensator***.

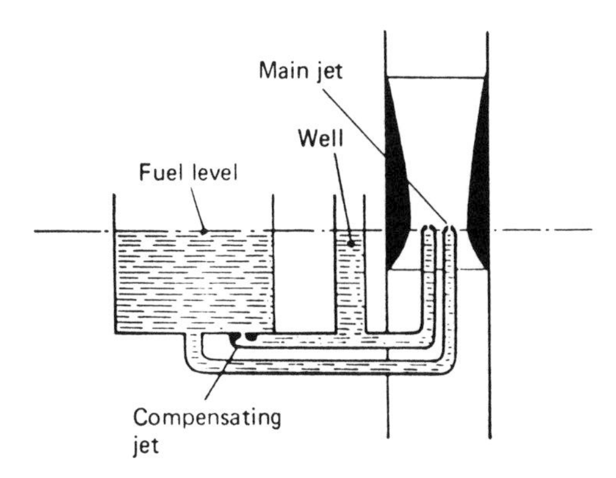

Figure C.8 Compensating jet in a simple carburetor.

compensator Device or system of linkages in a braking system to compensate for uneven wear, to ensure simultaneous operation of brakes on each axle, or to apportion braking effort. See also ***proportioner***.

compliance steer Steering input as a result of compliance in steering and suspension linkages.

composite trailer A ***semi-trailer*** effectively converted to a full trailer by addition of a ***converter dolly***. See Figure C.9.

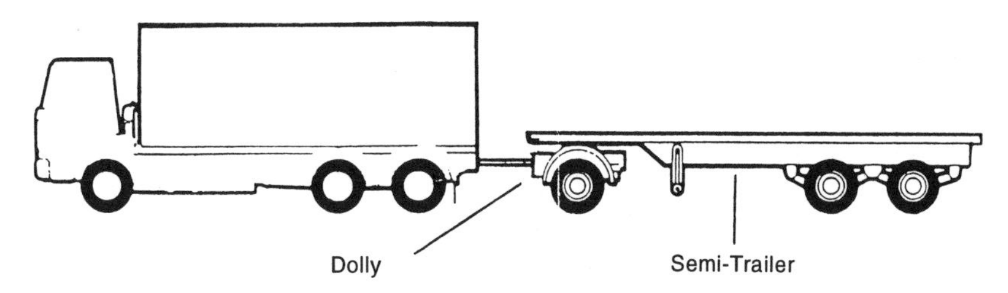

Figure C.9 Composite trailer with three-axle rigid towing vehicle.

compound carburetor Carburetor with more than one choke or mixing chamber per inlet port.

compressed natural gas Natural gas (principally methane) stored under pressure for use as an automotive fuel. Also ***CNG***.

compression (1) Of a solid material, subjecting to axial forces that would decrease the distance between points of load application. A coil spring is said to be in compression when carrying such a load, though the metal of the spring is actually in torsion. (2) The increase of pressure in an engine cylinder as the ***piston*** travels toward top dead center. The ***compression stroke***.

compression ignition engine Reciprocating engine in which the fuel charge is ignited spontaneously by the heat of compression. The term ***diesel engine*** is (mildly inaccurately) used to describe automotive compression ignition engines.

compression ratio In a reciprocating engine, the ratio of cylinder plus combustion chamber volume at the bottom of the ***stroke*** (at moment of greatest volume) to the volume at top of stroke, when the volume is least. Note that compression ratio is a volumetric ratio and not a pressure ratio. See Figure P.2.

compression ring The uppermost ***piston ring***.

compression stroke The stroke in a reciprocating engine in which the air or air/fuel charge is compressed prior to ***ignition***.

concept vehicle A vehicle produced for appraisal of a new or unusual design feature, or to test market response to an untried styling characteristic. Also ***concept car***.

conchoidal marks See ***beach marks***.

con-rod A ***connecting rod***. (Informal)

condenser (1) Device for condensing a liquid from its vapor. (2) An electrical ***capacitor***. (The term ***condenser*** is considered archaic in electrical and electronic circles.)

cone clutch Clutch in which the driving and driven elements engage on a conical surface. (Obsolete for most automotive power clutch applications, though used in the synchronizing mechanism of ***synchromesh gearboxes***.)

conicity The deviation from true symmetry about the wheel plane of a ***tire***, such that, instead of the tread profile forming an ideal cylinder, it forms a cone. In consequence the tire's path would naturally be circular and not straight.

conicity force The force necessary to correct the effect of conicity in a tire. The direction of this force is the same whether the tire (or vehicle) is moving forward or in reverse.

connecting rod Linkage that connects the crank to the piston in an engine or other reciprocating machine. See also ***crankpin end***; ***piston pin***. See Figure C.10.

connecting rod bearing An ambiguous term. Normally the ***big-end*** (or crankshaft end) bearing, though justifiably also the ***small-end*** or piston pin bearing. Usage discouraged. See Figure C.10.

constant depression carburetor Carburetor with variable section ***venturi***. Also ***variable choke carburetor*** or ***constant vacuum carburetor***.

constant mesh gearbox Transmission in which all forward gear pairs remain in mesh, the driving pair being engaged by a clutch mechanism, such as a ***dog clutch***, or by a ***synchromesh*** mechanism. See Figure C.11.

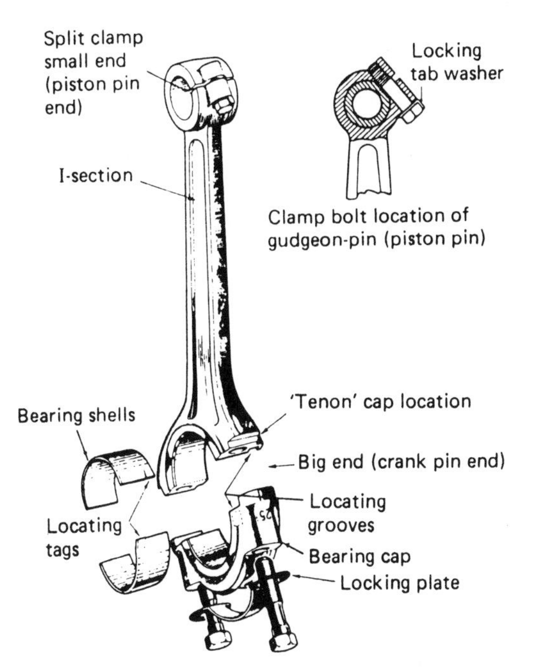

Figure C.10 Connecting rod with diagonally split big end (crankpin end).

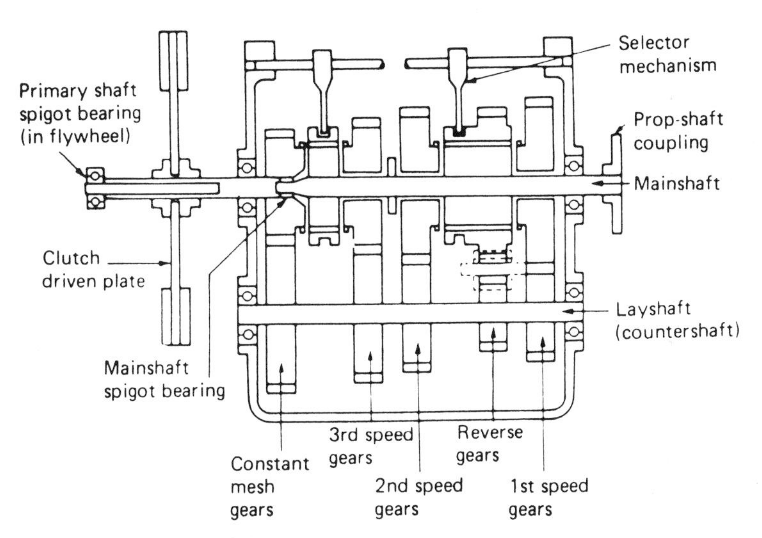

Figure C.11 Simplified diagram of a four-speed constant mesh gearbox (transmission).

constant velocity joint Universal joint in which the output shaft rotates at constant angular velocity with no cyclic variations, given a constant input shaft speed. See also ***Bendix-Tracta joint***; ***Bendix-Weiss joint***; ***Rzeppa joint***.

constant voltage control A ***voltage regulator***.

constantly variable transmission See ***continuously variable transmission***.

contact breaker Mechanical spring-loaded switch located in the ***distributor*** of an ***ignition system*** and in the low-tension circuit of a ***magneto***, and actuated by a rotating ***cam***, that makes or breaks the ignition circuit to deliver a spark to the ***spark plug***. See also ***breaker contact***; ***contact breaker points***; ***distributor***; ***electronic ignition***. See Figure I.1.

contact breaker points Hard metal cylindrical switch contacts whereby an ignition circuit is made and broken. Also ***points***. See Figure I.1.

contact center Intersection of contact line and normal projection of the ***spin axis*** onto road plane.

contact coupling Air line coupling with tapered plug and socket for connection of pressure and vacuum air brake systems. See also ***dummy coupling***; ***quick-detachable coupling***.

contact line Intersection of ***wheel plane*** and ***road plane***.

contact length Distance between leading and trailing edge of a tire footprint measured along ***wheel plane***.

contact maker Rotating switch for initiating an ignition spark by making the circuit. (Obsolete)

contact patch Area of contact of a tire with the ground. Also ***footprint***.

contact width Width of tire ***footprint*** measured at right angles to the ***wheel plane***.

container Enclosed rigid and secure box for conveying goods by road, rail or sea, and of a standard size and design.

continuous braking Braking of a combination of vehicles, such as a tractor unit and ***full trailer***, or an ***articulated vehicle***, where the operation of the brakes of the combination is controlled from one source, such as the effort applied by the driver to the brake pedal. See also ***automatic braking***; ***semi-continuous braking***.

continuous sampling Taking of samples continuously, as for ***exhaust analysis***, during which instrument readings are monitored.

continuous spray pump Fuel injector pump that provides a continuous rather than an intermittent spray, normally for ***fuel injection*** in spark ignition engines.

continuously variable transmission Transmission system in which the speed ratio of driving to driven elements is infinitely variable over the required working range, or (in less exact usage) is varied over a very large number of fixed ratios. Also ***constantly variable transmission***.

control arm Any lateral swinging arm of a suspension that controls the camber of a wheel, as for example the upper or lower arms of a double wishbone suspension.

control box (1) Container for the voltage and current ***regulators*** of an engine charging circuit. (Informal) (2) Any box or unit containing control circuits or presenting an operator with means of operating controls.

control lever Hand lever for selecting gears manually in an ***automatic transmission***. (Mainly US) Also ***stick-shift***. (Informal)

control ring An oil-scraper ***piston ring***.

controlled separation layer Separation of the airflow's boundary layer from a surface achieved by a discontinuity of body profile, as in a ***Kamm back*** vehicle.

converter (1) A ***catalytic converter***. Occasionally a ***thermal reactor***. (2) A ***torque converter***. (Informal)

converter dolly Stand-alone trailer with two or more wheels equipped with a ***fifth wheel*** to enable a ***semi-trailer*** to be towed by a rigid truck. See also ***composite trailer***. See Figure C.9.

convertible Passenger car with removable top or roof, enabling it to be converted to an open car. Also ***soft-top***; ***rag-top***. (US informal)

coolant Fluid, usually a liquid such as water or water/glycol mixture, used in a cooling system for an engine, compressor or other machine that requires cooling in operation.

cooling fan (1) Fan that draws cooling air through the ***radiator***. (2) Fan of a climatic cooling system. Also ***blower***. (Informal)

cooling system System of components devised for the cooling of an engine or other machinery, usually consisting of a radiator or other heat dissipator, a circulating pump, thermostat, coolant fluid and pipework.

copper A metal which, in its pure form, is used for its high electrical conductivity, particularly as wire or strip. Also available in alloy forms, notably with the addition of ***zinc***, as ***brass***, and with tin as ***bronze***. Chemical symbol Cu.

copper accelerated acetic salt spray A process for rapid assessment of a vehicle's resistance to corrosion. Also ***CASS***.

cornering force Force on a tire generated by the slip angle or distortion angle on cornering. Also ***lateral control force***; ***cornering resistance***. See also ***cornering resistance***.

cornering lamp Constantly illuminated lamp used in conjunction with the ***turn signal lamp*** system to supplement headlamp illumination in the direction of turn.

cornering resistance (1) That part of the total output of an engine that is used to overcome the additional resistance to motion of cornering. A contributing factor to total ***driving resistance***. (2) The longitudinal component of resistance due to cornering.

Corten-Dolan Theory Theory for evaluation of fatigue life on the basis of cumulative damage.

counter shaft (UK: layshaft) Shaft that runs parallel to the mainshaft in a ***gearbox***, and carries the ***pinion*** wheels. Also ***cluster gear***. See Figure G.1.

courtesy lamp An interior lamp for providing background lighting in a vehicle, particularly when it is not in motion.

cornering squeal Squeal produced by a free-rolling wheel when cornering.

coupling (1) Device for connecting together mechanical components, such as rotating shafts. (2) Means whereby a trailer is coupled to a tractor unit. Also ***coupling hook***.

cover The main outer body of a pneumatic ***tire***. See also ***carcass***.

cowl (1) Rigid covering or ducting, particularly when of nominally circular section. Also ***cowling***. (2) Front part of a vehicle cab or body directly below the base of the ***windshield*** and between the ***firewall*** or ***bulkhead*** and the ***instrument panel***.

crab tracked Having front and rear axle ***tracks*** considerably different, particularly where the rear track is narrower.

crabbing Vehicle movement resulting from axle misalignment such that the longitudinal axis is not in line with the vehicle's direction of motion.

cracking The breaking down of large molecules into smaller molecules in an oil refining process.

crank (1) Arm attached to a shaft and carrying a pin or handle or pedal parallel to the shaft. (2) To turn manually by means of a crank or starting handle. To crank an engine. (3) To turn an engine crankshaft by an external agency, as for example a cranking or ***starter motor***. (4) A ***crankshaft***. (Informal)

crank throw Radial distance from ***crankshaft*** axis to ***crankpin*** axis, equal to half the stroke.

crank web Radial arm, usually of flat section, that supports the ***crankpin*** on a ***crankshaft***.

crankcase Part of the structure of an engine that contains and supports the ***crankshaft*** and ***main bearings***. See also ***monobloc construction***.

crankcase compression Induction process in some smaller ***two-stroke*** engines, where the mixture charge is compressed in a sealed ***crankcase*** by the descending ***piston*** prior to passing to the ***combustion chamber*** by way of a ***transfer port***. See Figure T.11.

cranking enrichment Provision of excess fuel for cold starting as by ***choke*** or ***electronic fuel injection*** control.

cranking motor (UK: starter motor) Electric motor for starting an engine. See also ***Bendix drive***; ***overrunning clutch starter motor***; ***pre-engaged starter***.

cranking speed Rotational speed at which an engine is turned for starting.

cranking torque Torque required to turn a motor for starting.

crankpin Journal bearing of the crank of a ***crankshaft***, to which the ***connecting rod*** is attached.

crankpin end The lower, and larger, end of a ***connecting rod*** that bears on the ***crankpin***. The end may be separable to enclose a ***shell bearing*** and to facilitate assembly to the crank, or may, with a ***split crankshaft***, be in one piece, housing a rolling element bearing. Also ***big end***. (UK informal)

crankshaft The main power shaft of a reciprocating engine, comprising the cranks that impart reciprocating motion to the pistons by way of their ***crank throw*** or offset from the shaft axis, and the ***journals*** whereby it is located and supported by the crankcase ***main bearings***. See Figure E.1.

crankshaft end bearing (UK: big-end bearing) The bearing between ***connecting rod*** and ***crankshaft***. See Figure C.10.

crankshaft journal bearing Bearing between ***crankshaft*** and engine ***crankcase***. See Figure C.10.

crankshaft throw See ***stroke***; ***throw***.

crash gearbox (1) Gearbox or transmission with parallel driving and driven shafts in which geared pairs of straight spur gears are engaged by the axial sliding of straight spur wheels to engage with paired wheels on a ***layshaft*** or ***countershaft***. (2) Any change-speed gearbox or transmission without ***synchromesh***.

crawler (1) A slow large vehicle. (2) Very high reduction gear ratio, for use when climbing steep gradients or hauling heavy loads. (3) A ***track-laying vehicle***. See Figure C.12. See also ***Caterpillar track***.

crawler tractor Off-highway vehicle with ***traction*** provided by chain tracks rather than wheels.

crazing See ***checking***.

crew cab Vehicle cab, for example a municipal vehicle or fire appliance, extended to carry additional crew, usually by way of a second row of seats.

critical speed Speed at which some, usually untoward, phenomenon manifests itself, as for example the rotational speed at which a drive shaft ***whips*** or ***whirls***.

cross flow Particularly of a four-stroke engine, having the intake and exhaust ***manifolds*** on opposite sides of the cylinder block so that the gas flow is ostensibly across the cylinder.

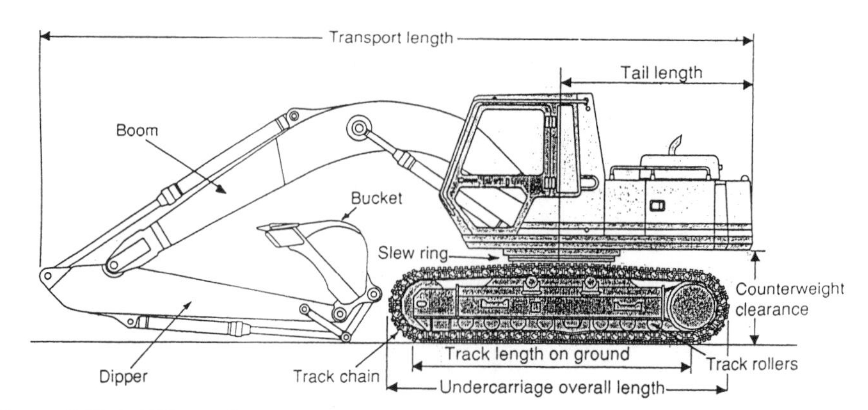

Figure C.12 A JCB crawler excavator with boom and dipper arm in transport position.

cross flow scavenging See ***cross scavenging***.

cross ply (US: bias ply) Form of ***tire*** construction in which the casing plies are laid diagonally and alternatively so that each ply lies at an included angle, usually of more than 40 degrees, to the adjacent ply. See also ***radial ply tire***. See Figure B.3.

cross scavenging Scavenging of a ***two-stroke engine*** by flow across the cylinder, a favorable flow pattern being induced by a wedge-shaped deflector on the piston crown. Also called ***cross flow scavenging***. See also ***loop scavenging***; ***reverse flow***; ***Schnuerrle system***; ***uniflow scavenging***.

cross-pins Cruciform component for carrying pinions or bearings, as for example the pinion carrier of a differential or the yoke connector of a ***Hooke's joint*** or ***Cardan joint***. (Informal)

crossbar Horizontal structural or stiffening member of a framework, as for example the upper member of a bicycle or motorcycle frame.

crude oil Naturally occurring oil prior to refining.

crown (1) The external road-contacting periphery of a ***tire***. (2) The top of a piston. See ***piston crown***.

crown wheel A bevel gear wheel in which the teeth are set around the periphery, giving the wheel the appearance of a crown. The larger wheel of a crown wheel and pinion bevel pair. See also ***differential***. See Figure D.5.

cruise control Automatic control that adjusts engine output, and selects gear in an automatic gearchange system, to maintain a constant preselected speed. Also ***automatic speed control***.

crush zone Part of a vehicle bodywork designed to absorb energy of structural collapse in collision thus reducing its transmission to the passenger compartment.

crystal chromatic light Lamp unit which is white or clear when unlit, but which gives off amber or other colored light, depending on type of filter, when lit.

cup wear Form of tire wear in which wear occurs periodically around the ***tread band*** circumference. Also ***cupping***.

curb rib Raised circumferential rib on ***sidewall*** of tire, to prevent damage to sidewall in contact with curb. Also ***kerbing rib***; ***sidewall rib***. See Figure B.3.

curb weight (UK: kerb weight) Weight of vehicle with fuel, lubricants and coolant, but without driver, passengers or payload.

cuff valve engine ***Sleeve valve*** engine in which the sleeves do not operate in sliding contact with the working pistons.

curing Chemical process of changing a material from liquid or paste form to a solid. See also ***drying***.

current regulator Electromechanical switch which opens when the current from the generator exceeds a certain value. See also ***control box***; ***voltage regulator***.

curtain-sider Commercial goods vehicle or trailer body with sides of a removable, flexible fabric secured over a framework and fastened at the base, which can be drawn back to allow side access to the load. See Figure C.13. See also ***Tautliner®***.

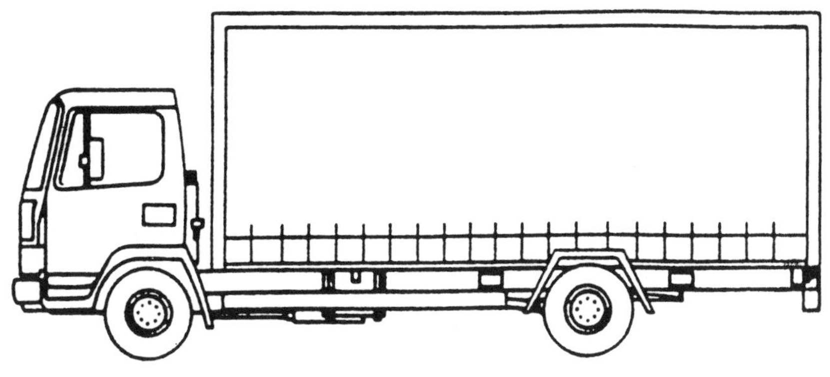

Figure C.13 Curtain-sider.

cushion drive Vehicle drive line incorporating a clutch with a ***cushion spring***.

cushion spring Flat annular spring which provides resilience between the friction linings of a ***clutch disc***, thereby cushioning the shock of engagement.

custom car Production car modified to personal tastes, but particularly one individually finished with elaborate decorative paintwork.

cut-in speed Rotational speed at which output voltage from a generator exceeds the voltage across the battery terminals, thus allowing the battery to be charged. See also ***cut-out***. Also ***cutting-in speed***.

cut-off frequency In acoustics, lowest frequency at which a specified material will effectively absorb all incident sound.

cut-out (1) Any device that halts the operation of a mechanical unit or electrical circuit. (2) Electromagnetic switch in control box that protects a ***dynamo*** against reverse current flow when battery potential exceeds dynamo voltage. See also ***cut-in speed***.

cutting-in speed See ***cut-in speed***.

CVT See ***continuously variable transmission***.

cyanoacrylates High-performance adhesives based on ***acrylic resins***, and capable of very quick setting at normal temperatures. Capable of bonding rubber, some plastics, and metals.

cycle life Of an electrical battery, the number of charge and discharge cycles that the battery can tolerate before appreciable degradation, often related to percentage depth of discharge.

cyclemotor Motor for propelling, or assisting the propulsion of, a pedal bicycle.

cyclone A centrifugal separator. See also ***filter***.

cylinder Cylindrical or tubular chamber in which the piston of a reciprocating engine or pump reciprocates. See also ***bore***, ***stroke***, ***cylinder block***. See Figure E.1.

cylinder block The part of an engine containing the cylinders. The cylinder block may also incorporate the water cooling jackets and provision for the ***valve gear***. See also ***monobloc construction***. See Figure E.1.

cylinder head Part of a reciprocating engine that seals or closes the upper ends of the cylinders. See also ***fixed head***; ***gasket***. See Figure C.14.

cylinder liner Thin-walled hard metal cylinder inserted into a ***cylinder block*** of an engine and in which the piston runs. See also ***dry sleeve***; ***wet sleeve***. See Figure A.2.

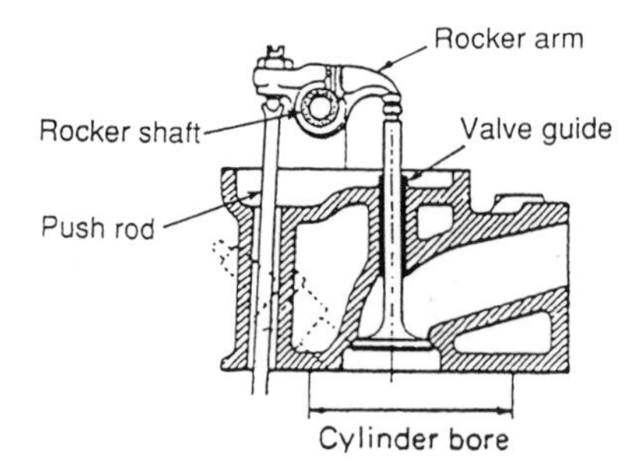

Figure C.14 Cylinder head with valve set in an offset bath tub.

D

D-post Structural support of the roof on a vehicle, and particularly a four-door sedan (saloon), against which the trailing edge of the rear door closes. See Figure B.4.

damper Device for dissipating energy of vibration, and hence for reducing vibration, as for example in an engine, ***camshaft*** drive or vehicle ***suspension***. See also ***shock absorber***; ***steering damper***; ***vibration damper***.

damper springs Springs set within a ***clutch plate*** to absorb shock loads arising from sudden or uneven engagement. Also ***torque cushion springs***.

damping Dissipation of energy in a vibrating system, usually by mechanical friction or fluid flow through an orifice.

damping slipper Damping device bearing on a ***chain and sprocket drive*** (and particularly an ***overhead camshaft*** drive) to minimize periodic oscillation or ***thrash*** of the chain.

dash See ***dashboard***.

dash panel (US: firewall) Bulkhead between passenger compartment and engine bay. See Figure B.4.

dashboard Interior panel beneath the ***windscreen*** or ***windshield***, on which instruments are mounted. Also ***dash***; ***fascia***.

day-cab Commercial vehicle cab without sleeping accommodation.

daylight opening Maximum unobstructed opening through any glass aperture. Also ***DLO***.

dc generator (UK: dynamo) Direct current rotating electrical generator. See also ***generator***.

decibel Logarithmic unit of sound level.

de Dion axle Combined system of power transmission and suspension in which the rear wheels of a vehicle are carried on a sprung ***dead axle*** or ***beam axle***, the final drive to each wheel being by ***cardan shafts*** from a chassis-mounted ***differential***. See Figure D.1.

de Dion joint A shaft joint allowing relative axial movement of the joined shaft. A ***plunging joint***. Also ***sliding-block joint***.

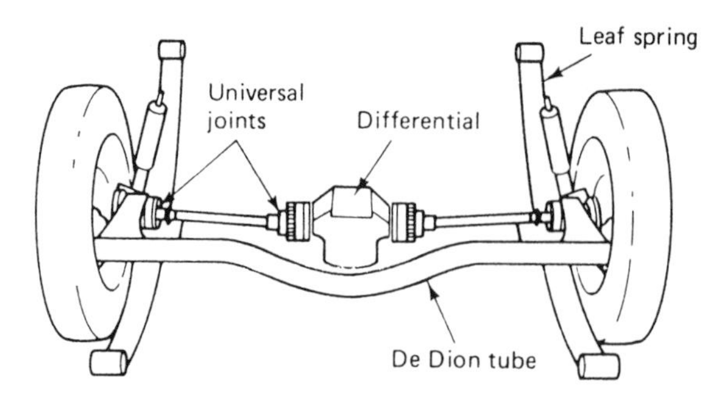

Figure D.1 One of the many configurations of the de Dion rear-wheel-drive system. Coil springs are often used as an alternative to leaf springs. The de Dion tube may also incorporate a sliding joint.

de Dion tube The offset ***dead axle***, sometimes incorporating a ***sliding joint***, of a de Dion rear suspension and ***final drive*** system. See also ***de Dion joint***.

de Normanville transmission Change-speed gearbox in which ratio changes are effected by applying hydraulically actuated ***band brakes*** or external ***shoe brakes*** to the outer drums of ***epicyclic*** gear units. Despite mechanical similarity to ***automatic*** and ***pre-selector*** gearboxes, the de Normanville is a manual shift transmission.

dead axle A non-driven axle, as for example the rear axle of a front-wheel-drive vehicle, or a ***de Dion axle***. See also ***beam axle***.

dead center The location of the piston of a reciprocating engine when at either extreme of its stroke. There are two such positions or "dead centers," one when the piston is at the top of its stroke (top or upper dead center) and one when the piston is at the bottom of its stroke (bottom or lower dead center). The intersection of the ***cylinder*** axis and ***crankpin*** axis indicates the dead centers only in engines in which the bore axis and crankshaft axis are in the same plane, which is not the case in engines with offset bores. Inner and outer dead center are alternative terms, though less often used. See also ***desaxé engine***.

decarbonize To remove carbonaceous deposits from the cylinders (and exhaust tract) of an engine. Also ***decoke***. (Informal)

deceleration valve Valve that allows extra air to flow into an intake manifold on deceleration to prevent backfiring. See also ***antibackfire valve***.

deck The floor of a vehicle, but particularly of a passenger vehicle such as a bus. See also ***cargo floor***; ***double-deck bus***.

decklid The lid of the ***trunk*** (US) or ***boot*** (UK) of a car.

decoke To ***decarbonize***. (Informal)

decompressor Valve, often manually operated, to allow free passage of ambient air to and from an engine cylinder to facilitate manual starting without turning the engine against compression. The decompressor is released before the ***compression stroke*** when adequate rotational speed has been reached.

deep cycling Repeated total discharging and recharging of an electrical storage battery.

deflection See ***static tire deflection***.

deflector crown piston See ***deflector piston***.

deflector head Two-stroke cylinder head of usually hemispherical shape to accommodate a deflector piston. See Figure T.11.

deflector piston Two-stroke piston configuration associated with ***cross scavenging***, in which the piston crown is in the form of a wedge-shaped gas deflector. See Figure T.11.

defogger Device or system for dispelling condensation on front or rear screen (shield), either by a flow of warm air or by an electric element within or attached to the glass.

defrost system Heater or system of heaters and fans for melting frost or ice on ***windscreen (windshield)*** or ***backlight***. See also ***defogger***. (UK)

defroster See ***defrost system***.

delay nozzle ***Pintle*** type injector nozzle that provides a ***pilot spray*** prior to the main injection of fuel.

Delco-Moraine brake Form of ***sliding caliper*** disc brake.

delivered air-fuel ratio The ratio of mass of air to the mass of fuel, delivered to, but not necessarily combusted within, the cylinder(s) of an engine. See also ***carburetor***; ***rich mixture***; ***stoichiometric; weak mixture***.

delivery van Light commercial vehicle equipped for mainly short-distance delivery of goods. Also ***delivery truck***. See also ***luton***.

deltic engine Opposed piston ***diesel engine*** with banks of three cylinders set in triangular configuration.

Demerit Rating System of rating engine deposits of combustion and lubrication, typically on a scale of 0 to 10, with 0 (zero) indicating complete cleanliness.

demister See ***defogger***.

demountable Free-standing rigid box structure or container for conveyance of goods, fitted with retracting legs to facilitate mounting on a specially adapted commercial vehicle which backs beneath the box body, whereupon the body can be lowered onto the vehicle's chassis bearers. A ***swap-body*** (also swop-body). See Figure D.2.

departure angle Maximum ***ramp angle*** that a vehicle can leave without fouling any part of the vehicle. See Figure L.3.

Figure D.2 A demountable freight system.

deposit induced runaway surface ignition Pre-ignition caused by progressive deposit formation. Also ***DIRSI***.

depression Static pressure less than ambient, as, for instance, in a venturi.

depth of discharge The proportion of total electrical storage capacity by which a battery is discharged, usually expressed as a percentage. Also ***DoD***.

desaxé engine Engine in which the crankshaft axis is offset from the cylinder axis, to increase the torque on the power stroke and possibly reduce side load on the piston/cylinder interface.

design H-Point See ***H-point***.

design weight Maximum weight at which a vehicle is designed to operate. This figure may be greater than the weight permitted by local or national regulations. See also ***plated weight***.

desmodromic valve Inlet or exhaust valve which opens and closes under positive cam action, sometimes with spring-assisted final seating.

desorption Loss or release of ***adsorbed*** materials from a substrate.

destination board Prominently displayed notice, usually alterable and illuminated, to advertize the route or destination of a bus or other passenger-carrying public service vehicle.

detergent Lubricant or fuel ***additive*** that reduces formation of carbonaceous deposits and other deposits in an engine's fuel system.

detergent-inhibitor Detergent additive to lubricants that also has ***anti-oxidant*** properties.

detonation Rapid and uncontrolled combustion. Detonation can occur in the cylinder of a spark ignition engine when operating on a fuel of inadequate octane rating, or with ***ignition timing*** too far advanced. Informally called ***pinging*** (US) and ***pinking*** (UK) in a spark ignition engine. See also ***diesel knock***; ***pre-ignition*** (which is not the same thing).

diagnostic testing Testing of a vehicle, or vehicle system, as for example an electrical system, engine, etc., to identify causes of malfunction or maladjustment. Generally used to describe automated or computerized workshop testing of production vehicles.

diagonal tire Tire in which the ***ply cords*** extend to the beads and are laid at alternate angles to the tread centerline. The term ***diagonal tire*** is more commonly used in the US, ***crossply*** in the UK, but terminology varies from manufacturer to manufacturer. The two types are constructionally similar. Also ***bias tire***. See also ***crossply***. See Figure B.3.

diaphragm Flexible disc or membrane which deflects under pneumatic or hydraulic pressure. In so doing, it may impart linear motion to a rod, as in an ***air brake*** or ***carburetor***.

diaphragm carburetor Floatless carburetor incorporating one or two diaphragms and capable of operating at any angle and normally unaffected by vibration and acceleration.

diaphragm chamber Air-tight metal chamber or drum containing the diaphragm and pushrod to apply the brakes in an ***air brake*** system. See also ***brake chamber***.

diaphragm clutch Clutch in which the ***pressure plate*** is maintained in contact with the ***friction plate*** by a ***diaphragm spring***. See Figure D.3.

diaphragm spring A disc-shaped metal spring, used particularly in clutches. The spring force is exerted by bending of fingers formed by radial slotting. See also ***clutch pressure plate***; ***clutch release bearing***; ***diaphragm spring ring***. See Figure D.3.

diaphragm spring clutch See ***diaphragm clutch***. See Figure D.3.

diaphragm spring ring Fulcrum ring about which a clutch ***diaphragm spring*** pivots when the clutch is actuated. See Figure D.3.

Diesel cycle ***Thermodynamic cycle*** in which air is compressed, heat added at constant pressure by injecting fuel into the compressed charge, the combusting mixture expanded to do work on the piston, and the products exhausted at completion of the cycle. After inventor and patentee Dr. Rudolf Diesel (1858-1913). See Figure D.4. See also ***adiabatic engine***; ***diesel engine***.

diesel electric drive Engine-transmission system in which a ***diesel engine*** drives an electric generator which, in turn, drives an electric motor. The system allows the engine to run at close to optimum speed, and eliminates the need for a change-speed gearbox. Generally used only on large industrial and off-highway vehicles, and railway locomotives.

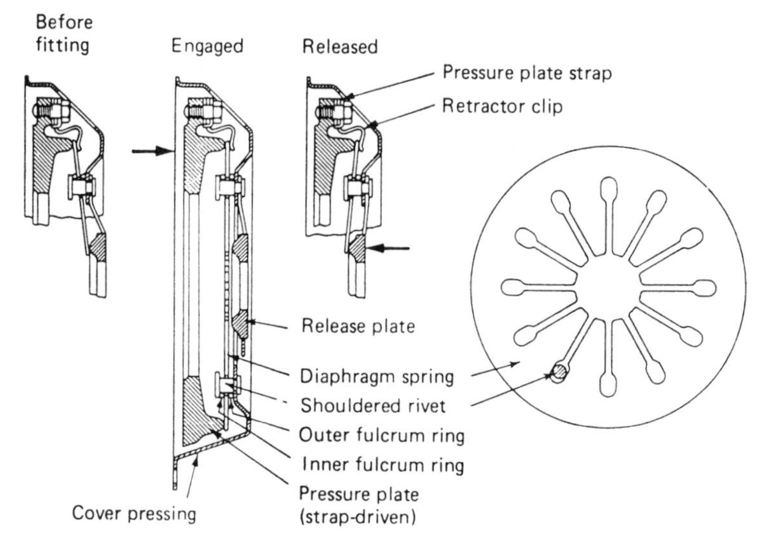

Figure D.3 Diaphragm clutch assembly and diaphragm spring.

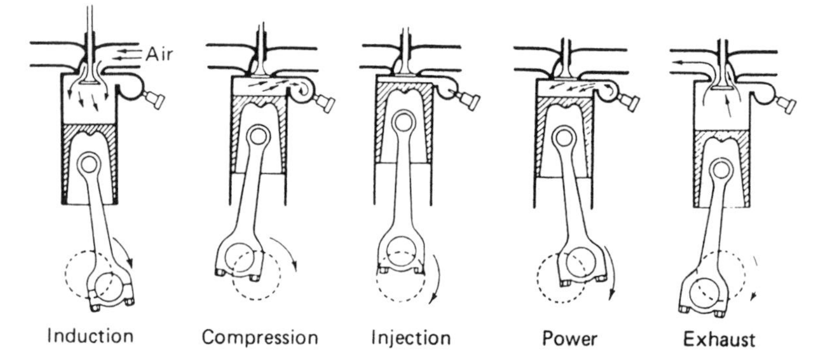

Figure D.4 The operating cycle of the practical high-speed diesel engine deviates from that of the ideal diesel cycle. In the conventional four strokes of the diesel cycle, "injection" and "power" would normally be counted as one stroke.

diesel engine (1) Reciprocating engine operating on the compression ignition ***Diesel cycle***, in which a charge of vaporised fuel is injected into the cylinder at completion of the compression stroke. (2) Informally used to describe any compression ignition engine. See Figure D.4. See also ***Cetane Number***; ***compression ignition engine***; ***Diesel cycle***; ***direct injection***; ***indirect injection***.

Diesel Index A measure of diesel fuel quality, derived from the aniline point test and related to ***Cetane Number***. The Diesel Index is generally superseded by the ***Cetane Index***.

diesel knock The noise made by the rapid pressure rise in certain types of CI engine, particularly at low speed and load. The effect is also encountered with fuels of lower ignition quality. See ***detonation*** (which is not strictly a form of detonation).

dieseling See ***auto-ignition***; ***run-on***.

diester Alcohol-derived component of ***synthetic lubricants***.

diff A driving axle ***differential*** gear unit. (Informal)

diff-lock See ***differential lock***. (Informal)

differential System of gears capable of dividing the input torque of one shaft between two output shafts where rotation at different speeds is likely to occur, as in cornering. Used as final drive of vehicles with two or more driven wheels. See also ***crown wheel***; ***limited-slip differential***; ***spur differential***. See Figure D.5.

differential angle Difference between bevel angle of poppet ***valve seat*** face and that of its seating, particularly in an IC engine.

differential cage (US: differential housing) Rotating housing, usually a casting, that is attached to the ***crown wheel*** and carries the final drive pinions in a ***bevel differential***. See Figure D.5.

differential case Case or housing of a differential, often forming a structural part of a live beam axle. Also ***differential carrier***. (US) See also ***banjo axle***. See Figures D.5 and H.2.

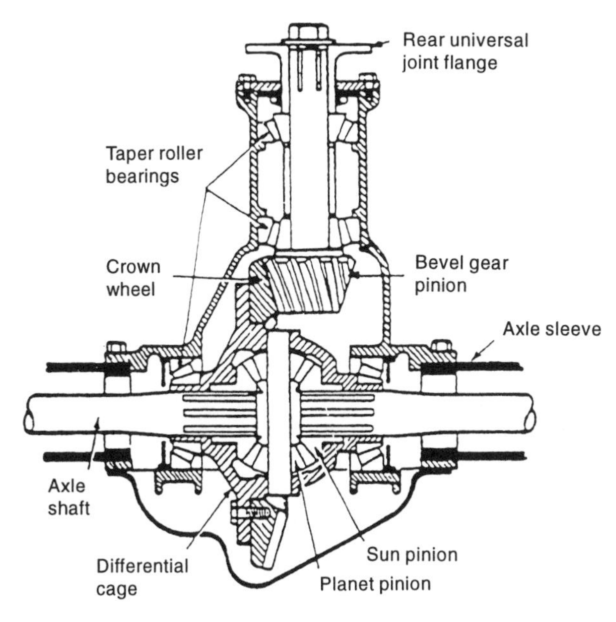

Figure D.5 Rear axle differential.

differential housing See ***differential case***.

differential lock Device for inhibiting the differential action of a differential, thus preventing ***wheel-slip*** of one wheel in slippery conditions. Also ***diff-lock***. (Informal) See also ***limited slip differential***.

differential protection valve See ***anti-compounding valve***.

diffuse field In acoustics, a space where sound pressure is equal in all directions.

diluent (1) Chemical, usually in liquid form, capable of thinning or reducing the viscosity of another substance, liquid or gaseous. (2) Chemical for adjusting the concentration of the active ingredient in an ***additive*** package in a ***fuel*** or ***lubricant***, or for making an ***additive*** that is more easily dissolved therein. (3) In painting and finishing, a substance capable of thinning paint or varnish, as for spraying. Also thinners. (Informal) See also ***solvent***.

dilution See ***oil dilution***.

dilution air In ***exhaust emission*** testing, ambient air which is passed through filters to stabilize the ***hydrocarbon*** concentration and dilute the vehicle exhaust.

dilution factor Arbitrary numerical representation of the concentration of ***carbon monoxide***, ***carbon dioxide*** and unburned ***hydrocarbons*** in exhaust gases, derived from the ***stoichiometric*** equation.

dilution ratio An index of the dilution by filtered air of an exhaust gas test sample to the gas concentration in the exhaust.

dim-dip Lighting system whereby the dipped or lowered beam of the ***headlamps*** has an alternative lower intensity setting to prevent annoying oncoming drivers.

dimensional stability The ability of a material, and particularly a synthetic material, to retain its shape and physical dimensions under the effects of environmental change.

dimmer switch (UK: dip switch, dipper switch) Electrical toggle switch by which ***headlamp main beam*** is extinguished and ***dipped beam*** switched on.

DIN rating Standard for measurement of engine performance specified by the Deutsche Institüt für Normung (DIN), characterized by the requirement that the test be conducted with an engine driving all normal ancillary machinery. See also ***SAE rating***.

dip switch See ***dimmer switch***.

dipped beam (US: lowered beam) Lowered headlamp beam for illuminating road when meeting other road users. Also ***lower beam*** or ***meeting beam***.

dipper Articulated extension of boom in earth-moving machinery. See Figure C.12.

dipstick (US: oil gage) Graduated rod to indicate oil level in engine ***oil pan (oil sump)*** or ***gearbox***. See Figure E.1.

direct drive Transmission drive mode in which engine and transmission shafts rotate at same speed, by-passing the reduction stages of the gearbox. This term is more common in marine than in automobile engineering.

direct injection Diesel engine injection system in which the fuel is injected directly into the engine cylinder. See also ***indirect injection***.

direction indicator (US: turn indicator; turn signal lamp) Flashing lamp to indicate direction a driver intends to turn, normally mounted at each corner of the vehicle, with optional or mandatory repeater lamps at side. See also ***flasher unit***.

directional control Control of a vehicle and quality of its response in steering.

directional stability Ability of a vehicle to maintain its course, or remain under normal steering control, while subjected to directionally disturbing influences such as cross-winds, camber changes, or braking on irregular surfaces.

directly controlled wheel In an ***anti-lock braking system***, a wheel in which the braking action results from the signal provided by a sensor in the same wheel. See also ***indirectly controlled wheel***.

director plate Multiple orifice plate fitted within a gasoline ***fuel injection*** nozzle to aid accurate spray formation.

disc brake Brake in which external friction pads are brought to bear on the faces of a disc, usually by the clamping action of a ***caliper***. See Figure D.6.

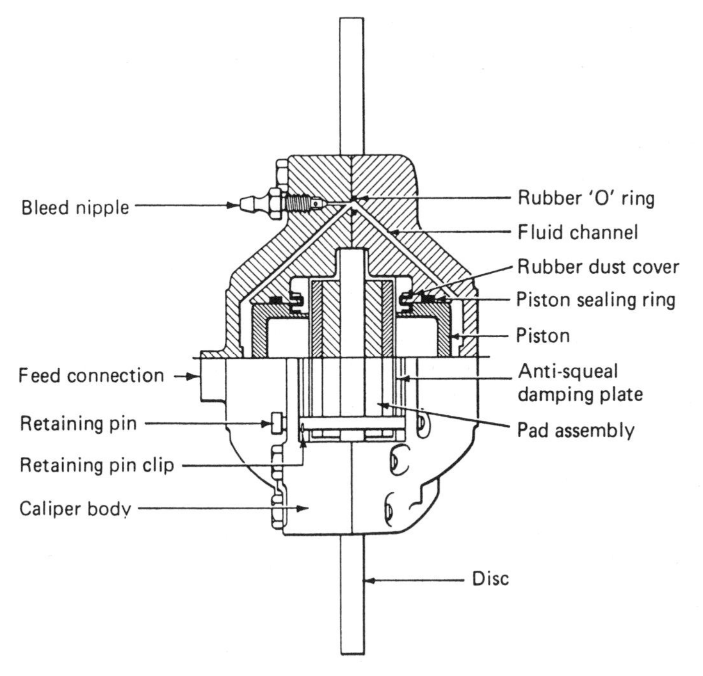

Figure D.6 Part section of disc brake assembly.

disc valve Rotating valve that provides open passage to a fluid by presenting an arcuate slot to an aperture, sometimes used to time the admission of mixture in a ***two-stroke*** engine. A form of ***rotary valve***.

disc wheel Wheel consisting of an inseparable ***wheel disc*** and ***rim***, but particularly when made from pressed steel.

discharge cycle The characteristics of the way in which an electrical battery is discharged, whether at high current, low current, or intermittently. Discharge cycle can influence the life of a battery, and other parameters. See also ***discharge rate***.

discharge orifice Hole or holes through which the fuel is injected in a fuel injector.

discharge rate The rate at which an electrical battery is discharged. A function of the current.

discharge tube Tube by which an emulsified air-fuel mixture enters the venturi of a ***fixed choke carburetor***.

dished Of a cylindrical form such as a piston, with a depressed end of shallow arcuate form. The opposite of ***domed***.

dispersant Engine oil ***additive*** which holds in ***suspension*** solid or liquid contaminants, thus reducing deposits and ***sludge*** deposition.

displacement The product of ***stroke***, and of cylinder ***bore*** and number of cylinders of an engine, representing the theoretical volume of air/fuel mixture that can be drawn into a cylinder with each ***induction stroke*** multiplied by the number of cylinders. See also ***capacity*** and ***swept volume***.

displacement factor Index of vehicle performance usually expressed as a product of engine displacement and ***axle ratio*** divided by the product of drive wheel ***rolling radius*** and ***gross weight***. (Mainly US usage)

dispersant oil Lubricating oil formulated to inhibit ***sludge*** formation.

distortion angle Angle between plane of wheel and direction of motion of wheel in a steered vehicle. Also ***slip angle***.

distributor Engine-driven rotary switch that switches the high-voltage ignition current to each ***spark plug*** in turn. See Figure I.1.

distributor cam A multi-lobed cam that actuates a ***contact breaker*** to initiate an electrical discharge to each cylinder in turn. See Figure I.1.

distributor cap The cover for the ***distributor***. This item usually incorporates the attachments and terminals of the ***spark plug*** leads and the high tension lead to the ***coil***, and is made of non-conductive material.

distributor pump High-pressure ***fuel pump*** that meters and delivers the fuel sequentially to the individual cylinders of multi-cylinder ***diesel engines***. See also ***in-line fuel injection***; ***injection pump***. See Figure D.7.

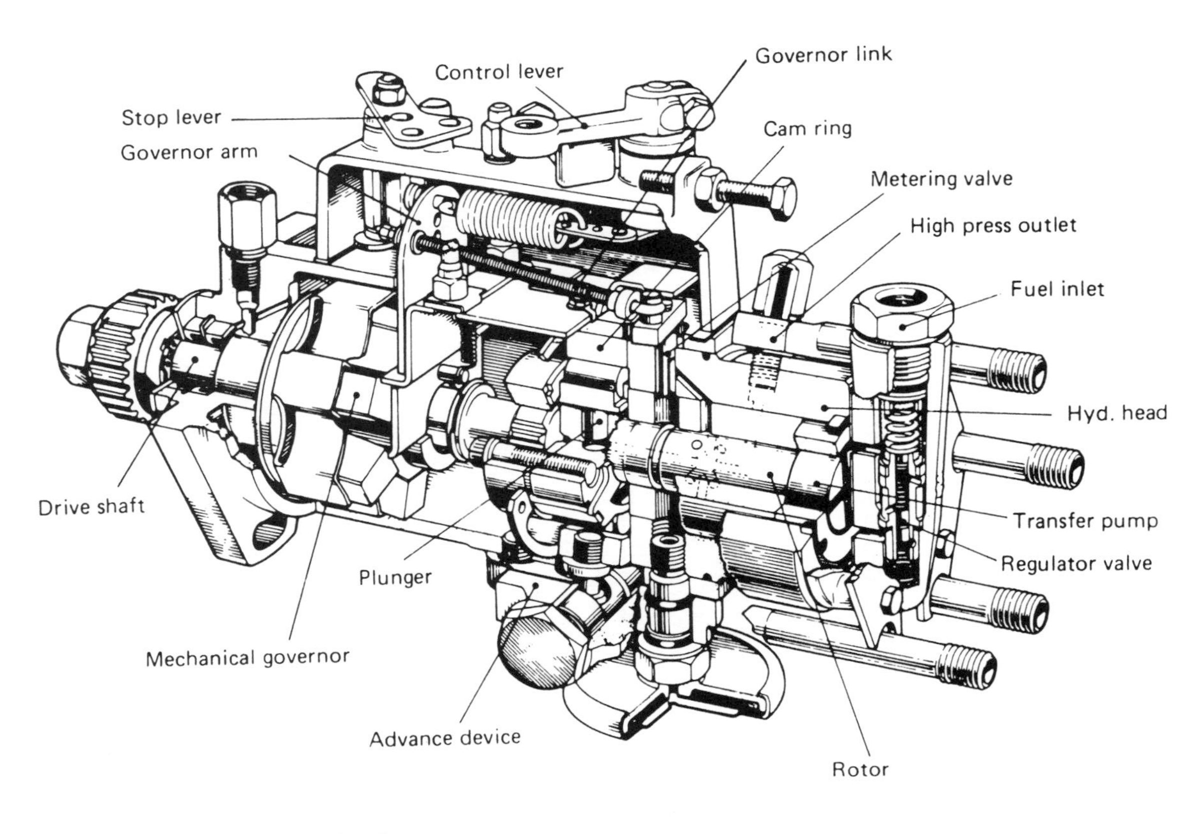

Figure D.7 The CAV DPA distributor pump.

distributorless ignition System of ignition in which the high tension (high voltage) electricity is switched electronically to each cylinder, rather than by an electromechanical rotary switch. The term is misleading, the distributor merely functioning in a different way. ***Electronic ignition*** is the preferred term. See also ***distributor***.

dive Nose-down pitching of a vehicle, as on braking. See also ***squat***.

divergent instability Instability in which a disturbance, as for example to direction or heading, leads to an increasing deviation without oscillation.

divided propeller shaft Shaft drive system between gearbox and final drive comprising two shafts with central bearing attached to chassis, suspension travel being facilitated by movement of the final shaft. Mainly employed to prevent ***whirling*** problems with long shafts on heavy vehicles. See also ***Hotchkiss drive***; ***torque tube transmission***.

divided system See ***split system braking***.

divided wheel Wheel consisting of two similar parts which are bolted together. Also ***split wheel***. See Figure W.5.

division-control multivibrator Electronic engine control circuit that determines the pulse characteristics of an ***electronic injection*** system from data feedback of engine speed and inlet air flow. See ***divided system***.

dog bone drive Drive, for transmitting rotary motion from one shaft to another, consisting of a ***connecting rod*** with bearing housings, often equal sized and engaging with eccentrics at each end, and thus shaped like a dog's bone. Occasionally used as an alternative to the chain or belt for driving an ***overhead camshaft***.

dog clutch Clutch which transmits power by engaging metal teeth or dogs. It allows only direct mechanical engagement or disengagement without slipping or progressive torque transmission.

dolly (1) Wheeled apparatus, often equipped with a ***fifth wheel***, for handling uncoupled ***semi-trailers***. (2) ***Landing legs***, usually equipped with a set of wheels, to facilitate handling a ***semi-trailer***. See also ***converter dolly***.

domed Raised to arcuate contour, as of a ***piston crown***. See also ***dished***.

donkey engine A small auxiliary engine, usually for driving ancillary equipment or services, for example when the main engine is stopped.

dope Fuel ***additive***, particularly of one of the ***aromatic*** hydrocarbons, used in competitive events where regulations permit. (Informal)

Doppler effect The effect of change of pitch from higher to lower as a vehicle approaches and passes.

double See ***double trailer***.

double-acting Of an engine or pump, where work is done on, or by, each side of the piston. Extremely rare in diesel engines, though common practice in steam.

double-barrel carburetor Carburetor in which two barrels share the same ***float chamber***. Typically used where ***intake manifold*** has two intake ports, and in competition engines.

double bottom Articulated tractive unit towing two ***semi-trailers***. See also ***double trailer***.

double cardan joint Constant velocity ***universal joint*** employing two ***Cardan*** or ***Hooke's joints*** in series. This type of joint accommodates limited axial misalignment, and approximates ***constant velocity*** operation whereas a single joint does not.

double-clutch See ***double-declutch***.

double-deck bus Bus with an upper passenger deck set above a lower deck, each deck being equipped with a gangway. A ***double-decker***. (UK informal)

double-declutch (US: double-clutch) Double operation of the clutch pedal to facilitate gear changing in a vehicle with a ***crash gearbox*** or ***sliding mesh*** gearbox.

double drive Vehicle having two driving axles.

double offset universal joint ***Constant velocity joint***, usually of the form in which caged balls transmit torque between inner and outer grooves, which can accommodate some axial misalignment as well as axial end movement or plunging. See also ***plunging point***; ***pot joint***.

double pivot steering Steering in which each ***steered wheel*** pivots about its own ***kingpin***—the normal system for almost all modern vehicles. See also ***Ackermann steering***; ***Jeantaud steering***; ***single pivot steering***.

double-reduction axle Heavy vehicle axle with two stages of reduction gearing between propeller shaft and final drive. Also ***two-speed axle***.

double trailer Articulated commercial vehicle also towing a ***full trailer***. Also ***double bottom***, ***double***, and ***turnpike double***. (US informal)

downdraft Carburetor through which the intake air flows vertically downward into the ***manifold***.

downpipe See ***head pipe***.

downshift (UK: change down) To select a lower gear.

downstream injection Gasoline ***fuel injection*** system in which fuel is injected at the downstream end of the inlet tract, often directed at the inlet valve. Also ***downstream spray***.

downstream spray See ***downstream injection***.

downtake pipe First exhaust pipe after ***exhaust manifold***. See also ***breeches pipe***. See Figure E.2.

downwash Downward component of air motion, particularly following a moving vehicle. See also ***upwash***; ***vortex pair***.

dozer See ***bulldozer***. (Informal)

dozer blade A massive, usually concave, and vertically inclined blade for stripping, leveling and shallow digging in earthmoving and construction work.

drag (1) Air resistance. The external aerodynamic friction forces that resist movement of a vehicle. (2) Resistance to movement caused by oil viscosity. (3) Acceleration competition between vehicles, and particularly between vehicles built to achieve high acceleration over short distances. See also ***dragster***.

drag coefficient The aerodynamic drag of a vehicle per unit cross-sectional area, as a non-dimensional quantity. An indication of aerodynamic efficiency of a road vehicle. Note that drag coefficient in flight context is a different expression. See also ***profile drag***.

drag link The link in a steering system that connects the ***drop arm*** or ***Pitman arm*** to the steering arm. Also ***steering side tube***. See Figure S.8.

dragfoiler See ***air shield***.

dragster Competition or record vehicle for straight-line sprinting or drag racing.

drain plug Plug, often hexagon-headed and threaded, removal of which allows a fluid to be drained from a reservoir or tank such as an engine ***oil pan*** or ***sump***. See Figure E.1.

drain valve Valve by which condensation is removed from the reservoirs of an ***air brake*** system.

drainback passage Passage through which a recirculating fluid, such as a lubricating oil, returns to its source.

drawbar Horizontally hinged rigid bar or ***A-frame*** by which a ***full trailer*** is towed and steered, the forward pair of wheels being steered by the drawbar assembly. Sometimes draw-bar.

drawbar combination Rigid commercial vehicle towing a drawbar trailer.

drawbar pull Tractive force exerted at the drawbar of a towing vehicle, sometimes expressed as factored power per unit speed, which gives a value in units of force.

drawbar rig See ***drawbar combination***. (Informal)

drawbar test Method for evaluating resistance and other parameters of a vehicle by towing with a suitably instrumented ***drawbar***.

drawbar trailer Full trailer, with axles at front and rear, towed by means of a ***drawbar*** by which the front wheels are steered.

drawing vehicle Vehicle equipped for and engaging in towing. (Mainly UK official usage)

dray Vehicle for conveyance of agricultural produce or drink. A brewer's dray. A ***low-loader***. (US informal)

drift Controlled lateral slide on a turn.

drip molding Narrow reflexed guttering attached to longitudinal roof edges. See Figure B.4.

drive-by-wire Control of engine or other vehicle functions by electronic rather than mechanical means. See also ***ABS***; ***cruise control***.

drive line (US: drive train) (1) Transmission system from engine output shaft to driven road wheels. (2) The term drive line sometimes encompasses the engine. See also ***powertrain***.

drive shaft (1) Shaft or shafts by which power is transmitted to a rear axle ***differential***. (2) Any rotating shaft by which power is transmitted. See also ***Hotchkiss drive***; ***propeller shaft***. See Figure H.2.

drive shaft ringing Acoustic phenomenon, usually due to excitation of the harmonics of the drive shaft, particularly if of plain steel tubing.

drive train See ***drive line***.

driveability Of an engine or vehicle, exhibiting ease of control, particularly of engine torque and low-speed operation. See also ***flexible***.

drivebox A ***gearbox*** or ***transmission***. (US informal)

driven disc See ***driven plate***.

driven plate Disc-shaped ***clutch*** element with annular friction lining to which torque is transmitted from the engine, and which transmits that torque through splines to a ***gearbox*** or other driven shaft. Also called ***clutch disc***, ***driven disc*** and ***friction plate***. See also ***pressure plate***. See Figure C.4.

driver (1) The person controlling a vehicle. (2) Car for everyday use. (US informal)

driver's field of view See ***field of visibility***.

drive shaft tunnel (UK: transmission tunnel) Longitudinal raised tunnel-shaped section along center of ***floor-pan*** to house ***drive shaft*** (propeller shaft).

driving axle Axle capable of transmitting power by way of a ***differential*** or other transmission arrangement. A ***live axle***. See also ***double reduction axle***; ***driven axle***; ***single reduction axle***.

driving beam See ***main beam***; ***upper beam***.

driving cycle See ***test cycle***.

driving lamp Lamp fitted to the front of a vehicle to illuminate the road in conditions of poor visibility and during hours of darkness. In some countries, driving lamps must be in constant use regardless of conditions of natural light. Also ***headlamp***.

driving mirror Mirror providing a driver with rearward vision, particularly when mounted within a vehicle. See also ***wing mirror***.

driving resistance The sum total of all the resistances to motion that must be overcome by a vehicle's engine, including ***rolling resistance***, acceleration, cornering, climbing and aerodynamic loads. Also ***total driving resistance*** (TDR). The term is not in universal use.

driving wheel A ***steering wheel***. (UK archaic)

dromedary Truck equipped with ***fifth wheel*** for towing a ***semi-trailer***. (US informal)

drop arm (US: pitman arm) Lever or arm that translates the rotary output of a ***steering box*** to the linear movement of a ***drag link***. Also ***steering arm***. See Figure S.8.

drop axle Dead axle of which the main portion of the beam is lower than the axis of the wheels. See also ***drop-center axle***. See Figure S.8.

drop box See ***drop gear***.

drop-center axle Dead axle, the center part of which is lowered, usually to give clearance to a ***drive shaft***. See also ***drop axle***.

drop-center rim Wheel rim with smaller radius center part to facilitate tire changing.

drop-frame trailer (UK: low loader) A ***flatbed trailer*** with platform raised only to clear the axles. See also ***step-frame trailer***.

drop gear Gearbox with output axis lower than input axis, used particularly on commercial vehicles where the engine is mounted considerably higher than the drive axles. See also ***transfer gear***.

drop-head Passenger car with collapsible fabric roof. (UK informal)

drop point Temperature at which oil commences to drip from a ***grease***.

dropped axle ***Beam axle*** in which the inner portion is significantly lower than the wheel axis. Also ***drop axle***.

dropside flat Flatbed truck or lorry with horizontally hinged sides that may be dropped for loading and unloading. See also ***platform body***.

dropwell Lowered portion of a vehicle floor.

drum brake Brake in which friction blocks called ***brake shoes*** lined with friction material are brought to bear on the periphery of a drum or cylinder. In most vehicle applications the ***shoes*** are brought into contact with the inner periphery of the drum. See also ***internal expanding brake***. See Figure D.8.

dry charged battery Lead acid battery with charged plates but containing no electrolyte. A frequent state for storage prior to sale.

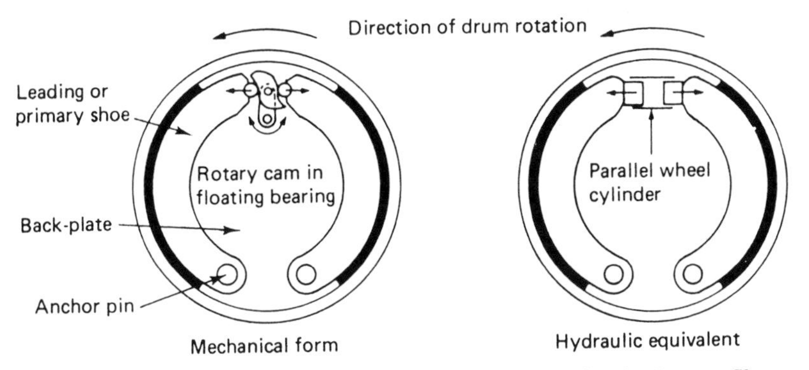

Figure D.8 Floating cam drum brake in mechanical and hydraulic forms. Shoe-to-shoe springs are omitted for clarity.

dry clutch Clutch which operates in air rather than in a bath of oil or other liquid.

dry lubrication Condition when rubbing surfaces have no lubricating liquid between them. See also ***boundary lubrication***.

dry sleeve (UK: dry liner) Hard metal engine ***cylinder liner*** that is not exposed to contact with cooling water.

dry sump Of an engine, when main lubrication is supplied from a remote reservoir, the ***sump*** or ***oil pan*** (if incorporated) containing no oil while the engine is operating. A common feature of motorcycle and competition car engines.

dry weight Weight of an unladen commercial vehicle without fuel, water and oil, but subject to manufacturer's interpretation. See also ***kerb weight***; ***tare weight***.

drying Conversion of a liquid or paste, such as a paint or adhesive, by evaporation and/or by atmospheric oxidation of components such as drying oils, to render it dry to the touch.

dual beam headlamp Single headlamp unit that provides both main (upper) and dipped (lower) beams. See also ***single beam headlamp***.

dual bed converter See ***three-way converter***.

dual control car Passenger car with duplicated controls for emergency use by a driving instructor.

dual-drive tandem Tandem axle in which both axles are driven or live.

dual fuel engine Engine capable of running on two distinct types of fuel, such as a gas and liquid fuel.

dual level system Vehicle light system that incorporates lamps which can be switched between high and low intensity, particularly ***stop lamps*** and ***direction indicators / turn indicators***.

dual overhead camshaft (UK: twin overhead camshaft) Arrangement of two overhead camshafts per bank of cylinders in an engine. (Mainly US usage) See also ***twin camshaft***.

dual-purpose vehicle Vehicle for carriage or passengers and goods or other burden, and fitted with a permanent roof.

dual spacing Lateral distance from ***tire*** centerline to the centerline of a ***dual tire*** arrangement. ***Track*** (UK) or ***tread*** (US) of a ***dual tire*** axle as variously defined.

dual tire Having two wheels closely coupled per side on one axle.

dual venturi carburetor See ***twin choke***.

dual-wheel Front wheel configuration, as on an adapted ***agricultural tractor***, in which wheels are mounted on ***stub axles*** set to give extreme ***camber*** and therefore minimum track.

Dubonnet suspension Steering-suspension arrangement in which a ***beam axle*** carrying kingpins at its ends is rigidly attached to the vehicle. ***Steering arms*** from the kingpins carry swinging ***suspension arms***, from which the wheels are mounted by ***stub axles***.

ductboards Removable raised or hollow boards which support a cargo on the ***cargo floor*** of a commercial vehicle and allow circulation of air for ventilation and drainage. Also ***duckboards***.

dummy coupling Air line sealing or blanking attachment for use on air brake lines of a towing vehicle when no trailer is attached. See also ***contact coupling***.

dump body Tilting body of a ***dump truck*** or tipper truck. See Figure R.6.

dump trailer ***Semi-trailer*** of ***full trailer*** equipped to discharge its cargo by tipping.

dump truck (1) A tipper truck. (2) An all-terrain construction vehicle with tipping hopper for earthmoving. A ***dumper***. (Informal) See also ***shuttledumper***; ***side dump truck***. See Figure R.6.

dump valve (1) Valve that operates relay emergency valve of a trailer from the tractor unit in the event of failure of a trailer line. (2) Any valve that releases the pressure in a pressurized system or container, in effect, dumping unwanted pressurized gas or air. A ***pressure relief valve***.

dumper A ***dump truck***. (Informal)

duo duplex brake See ***wedge-operated brake***.

duo servo brake Brake in which primary and secondary ***shoes*** are linked together so that there is only one abutment and not one for each shoe.

duplex chain Roller chain with two parallel sets of rollers, often used as an engine ***camshaft*** drive. See also ***timing chain***.

dwell angle Angular period of closure of ***contact breakers*** in a ***distributor***.

Dyer drive Positive engagement ***starter motor*** or cranking motor, usually for heavy-duty use.

dynamic balancing Balancing of components in rotation rather than at rest, particularly of wheels and engine crankshafts.

dynamic seal Seal between two surfaces or components in relative motion. See also ***reciprocating seal***; ***rotating shaft seal***.

dynamic viscosity Viscosity as measured by a rotating mechanical apparatus rather than by gravity flow. Sometimes known as ***absolute viscosity***. See also ***kinematic viscosity***.

dynamic supercharging Pressure charging of an engine using the kinetic energy of the induction air or the resonant properties of the inlet tract rather than the compression of the air by a mechanical device such as a ***turbocharger***. See also ***ram air induction***; ***tuned intake pressure charging***.

dynamo (US: dc generator) Direct current rotating electrical generator. See also ***generator***.

dynamotor A combined ***generator*** and ***starter motor***. (Obsolete on road vehicles)

dyno A ***dynamometer***. (Informal)

E

earth (US: ground) Return electrical circuit.

ECE Cycle Standard European vehicle test cycle. See also ***Euromix Cycle***.

ECE test Test of engine emissions quality in which diluted exhaust samples are collected in one sample bag.

eccentric (1) Any circular rotating element with an off-center axis. (2) The "crankshaft" of a ***Wankel engine*** and certain rotary pumps.

ECM See ***electronic control module***.

economizer Device that regulates the flow of fuel to a ***carburetor***, particularly at maximum demand, rarely fitted as original equipment.

eddy-current retarder See ***electric retarder***.

Edwardian car Car of period between Veteran and Vintage, including the reign of King Edward VII of England (1901-1910).

effective rolling radius The radius of a rigid hypothetical wheel with zero slip that, on rotation at the angular velocity of the actual wheel, would give to the vehicle its actual linear velocity. This radius is always smaller than the actual radius to the undistorted tire periphery, particularly with ***radial ply tires***.

effective static deflection (1) Deflection of a suspension system at a stated static load. (2) Static load of a loaded ***suspension*** system divided by the ***spring rate*** of the system at that load. See also ***static tire deflection***.

EGR See ***exhaust gas recirculation***.

eight-wheeler Four-axled rigid truck with two steering axles and usually two driven axles. (Mainly UK usage) See Figure R.6.

elastomer A material of large molecular structure, usually a ***synthetic rubber*** or plastic, with elastic mechanical properties.

elastohydrodynamic lubrication Lubrication in which the load is such that the lubricated surfaces deform, and in so doing enlarge the contact area.

electric brake (1) ***Service brake*** operated electromechanically rather than by hydraulic or mechanical means. (2) An ***electric retarder***.

electric retarder Rotating electromagnetic ***transmission brake***, effective only when vehicle is in motion.

electric vehicle Vehicle propelled by electric motor, drawing its current either from storage batteries or from overhead cables. Also ***EV***. See also ***trolley bus (trolleybus)***.

electrically heated catalyst Exhaust ***catalytic converter*** with facilities for electrically pre-heating to reduce ***light-off time***. Also ***EHC***. See also ***burner-heated catalyst***.

electrodynamic retarder See ***electric retarder***.

electromagnetic compatibility Extent to which vehicle electrical system is affected by external electromagnetic fields.

electronic control module Semiconductor unit for controlling ***ignition timing*** and other parameters in an engine management system. Also ***ECM***.

electronic ignition Ignition system in which switching semi-conductors make and break the low tension circuit. Also ***breakerless ignition***. (Informal) See also ***electronic triggering***.

electronic regulator Voltage regulator in which generated voltage is sensed and controlled by a Zener diode or other semi-conducting device. See also ***control box***; ***voltage regulator***.

electronic triggering Breaking of an ignition circuit by an electronic switch rather than by a mechanical (***contact breaker***) system.

electro-rheological Of a fluid, one whose flow characteristics can be modified by an electric or electromagnetic field. See also ***rheology***.

Elliot axle Arrangement of axle whereby the axle beam terminates in a yoke or fork-end which holds the ***kingpin***, the axle pivoting on an eye-end within the yoke. See also ***Lemoine***; ***reversed Elliot***.

elliptical spring Spring comprising two ***semi-elliptical springs*** shackled back-to-back. The resultant form is more accurately described as lenticular than elliptical. Also ***full elliptic***.

EMC See ***electromagnetic compatibility***.

emergency brake system Any brake system capable of bringing a vehicle to rest in the event of failure of the main braking system.

emission Any gas, vapor or particulate loss to atmosphere. See also ***carbon monoxide***; ***catalytic converter***; ***evaporative emissions***; ***exhaust emissions***; ***nitrogen oxide***; ***unburned hydrocarbons***.

emission control (1) Regulation, and by inference, reduction of toxic or pollutant content of a vehicle's ***exhaust gases***. (2) A device, such as a ***catalytic reactor***, for reducing pollution from an exhaust.

emulsion (1) Mixture of fine droplets of one fluid dispersed in another fluid. (2) Partially vaporised and heavily enriched fuel-air mixture, prior to being introduced into the main ***venturi*** of a carburetor. See also ***emulsion tube***.

emulsion block See ***emulsion tube***.

emulsion tube Combined main and compensating jet tubes in a ***carburetor*** with provision for drawing air into the fuel flow to create an emulsion at higher engine speeds, thus preventing over-richness and improving fuel distribution. Also ***emulsion block***.

end float Longitudinal play in a shaft, intentional or otherwise. See also ***play***.

end yoke Cusp-shaped termination as of a ***Cardan shaft*** or ***Elliot axle***.

energy absorber Device for absorbing energy. The term is mainly used for a device that absorbs the energy of impact, as for example an energy absorbing ***steering column***.

energy balance Quantitative audit of energy supplied to an engine (by fuel) balanced by the energy provided (and lost) by the engine as useful power, heat losses and exhaust gas losses, etc.

energy density The electrical storage capacity of a battery per unit volume of the battery, usually expressed as Watt-hours per litre. See also ***power density***.

energy transfer System of ignition, similar to the ***magneto***, with rotors operating within external coils. Mainly used on motorcycles.

engage To bring about mechanical continuity, as in engaging a clutch or a gear in a ***change-speed gearbox***. See also ***select***.

engine The main power unit or motor of a vehicle, converting the energy of a liquid or gas fuel into mechanical energy. Motor and engine are usually synonymous in vehicle terminology when referring to a conventional internal combustion engine. The propulsive unit of an electric vehicle is generally an ***electric motor***, while an external combustion engine, for example one running on the ***Stirling cycle***, may be referred to as a Stirling machine. See Figure E.1.

engine bay The space within a vehicle that accommodates the engine. See also ***bonnet***; ***hood***.

engine brake Auxiliary brake which uses the compression of inducted air by the engine's pistons as a means of absorbing energy. This system of auxiliary braking normally requires the overriding of the operating timing of the exhaust valves. Also called ***Jake brake***, after manufacturer's name. See also ***retarder***.

engine hour meter Instrument which automatically records the time for which an engine has run, as a guide to maintenance and service.

engine identification number (EIN) Manufacturer's coded reference or identification number unique to each engine. See also ***vehicle identification number (VIN)***.

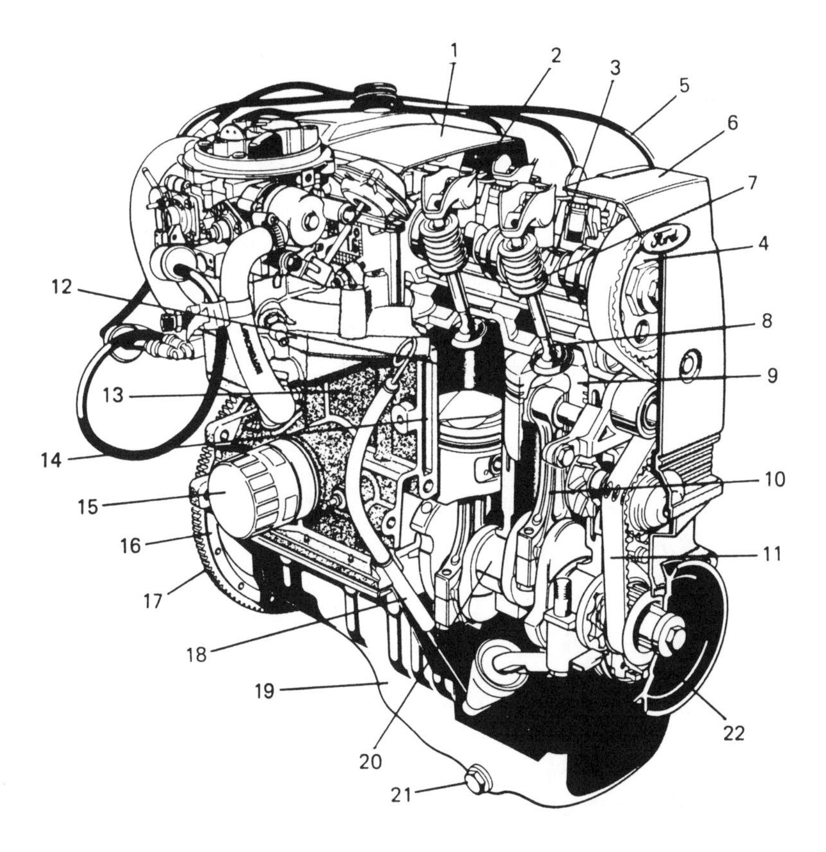

1 – Rocker arm cover (rocker box)
2 – Valve rocker arm
3 – Camshaft
4 – Camshaft drive
5 – Spark plug lead
6 – Timing belt cover
7 – Valve spring
8 – Poppet valve
9 – Piston
10 – Connecting rod
11 – Timing belt
12 – Combustion chamber
13 – Cylinder block
14 – Water jacket
15 – Oil filter
16 – Flywheel
17 – Ring gear
18 – Dip stick (oil gage)
19 – Oil pan/sump
20 – Crankshaft
21 – Drain plug
22 – V-belt pulley

Figure E.1 A Ford four-cylinder overhead-camshaft engine.

engine management system Arrangement of microprocessor-controlled electromechanical devices for controlling a vehicle engine.

engine map Three-dimensional graphic representation of engine control parameters, for example angle of ignition advance plotted against bases of throttle opening and engine speed.

engine mounting (1) Attachment points of an engine to a chassis or vehicle structure. (2) The means whereby an engine is supported. See ***anti-vibration mountings***.

engineering plant Special-purpose vehicle or trailer not for carrying goods or passengers, for example earthmoving equipment or a mobile crane.

enrichment In context of carburetion, increasing the proportion of fuel to air.

entrainment Dispersion of a liquid or gas in a fluid without solution.

EP See ***extreme pressure***.

EP additive Additive formulated to provide lubrication under ***extreme pressure*** conditions, as in ***hypoid drives***. See also ***EP lubricant***; ***extreme pressure***.

EP lubricant ***Extreme pressure*** lubricant for use in high-performance geared systems.

EPA Highway Cycle Standard test cycle issued by US Environmental Protection Administration.

epicyclic gearbox (US: planetary transmission) Gear system in which small pinion wheels run between an internally toothed annular wheel and a central externally toothed wheel, often called a ***sun wheel***. Input and output can be between any two of the three gears or gearsets, with one gear element constrained against rotation. The epicyclic principle is used in many ***automatic*** and ***pre-selector gearboxes*** and ***overdrives***.

epitrochoidal engine Rotary engine in which a section of the chamber forms an epitrochoid. The ***Wankel engine*** is an epitrochoidal engine. See Figure W.1.

epoxide Synthetic resin used as an adhesive, as a base for high-duty paints and finishes, and as a generally higher-performance alternative to polyester resins in the molding of glass fibre and other filled products. Available in two part and heat-curing formulations.

equalizer beam Pivoted beam joining the fore and aft springs of an interactive (reactive) ***tandem axle*** suspension. Also ***equalizer***. See also ***balance beam***; ***four-spring suspension***; ***walking beam***. See Figure T.1.

equatorial line Circle of intersection of ***tire tread*** surface of an unloaded tire with the ***wheel plane***.

equivalent braking force Ratio (usually expressed as a percentage) of the ***total braking force*** of a vehicle to the gross weight. (SAE definition)

ergonomic Designed with consideration for the comfort or controlling efficiency of a passenger or driver. For example, an ergonomic seat or control panel. See also ***hand-reach envelope***.

ergonomics In the context of vehicles, the (study of the) relationship between the human body and the vehicle or its control.

estate car (US: station wagon) Passenger car with extended constant height body fitted with tailgate or rear doors to facilitate access and provide stowage for bulky items. A ***shooting-brake***. (UK archaic)

ESV See ***experimental safety vehicle***.

ethanol Ethyl-alcohol as an ***alternative fuel*** or as a component of ***gasoline***. It is not classed as an ***additive***.

ethyl alcohol Fermentation product of starches and sugars used, among other things, as an alternative fuel or as a component of ***gasoline***. It can also be made synthetically. Also ***ethanol***; ***alcohol***. (Informal)

ethylene acrylic ***Elastomeric*** material with resistance to attack by oils at elevated temperatures.

ethylene glycol Chemical basis of anti-freeze.

ethylene methyl acrylate Base material of certain high-duty automotive ***elastomers***.

ethylene oxide-epichlorohydrin ***Elastomer*** which retains its flexibility at higher temperatures.

Euro 1 European exhaust emission legislation introduced in 1993.

Euro 2 European exhaust emission legislation introduced in 1996.

Euromix cycle Standard vehicle ***test cycle*** that simulates driving partly in urban conditions and partly on the open road.

evaporation losses See ***evaporative emissions***.

evaporative emissions Fuel vapors which escape to atmosphere by evaporation.

excavator Construction vehicle for earth-moving. See Figure C.12.

excess air factor Factor by which the ***air-fuel ratio*** of an inducted mixture exceeds that of the ***stoichiometric*** mixture, expressed by (trapped air-fuel ratio)/(stoichiometric ratio).

excess fuel device Device to provide an increased amount of fuel to an engine, as for starting, for example a ***choke***.

exhaust analysis Quantitative measurement and presentation of the constituents of an engine's exhaust gases, vapors and particulates.

exhaust back pressure Resistance that impedes the flow of the exhaust gases from engine to atmosphere, caused by the friction and other restricting factors in the exhaust system.

exhaust brake System of retarding a vehicle by constricting the flow of engine exhaust gases, and thereby increasing the retarding effect of the engine on overrun. See also ***engine brake***.

exhaust calorimeter Device for measuring the heat content of an engine's exhaust.

exhaust emissions Substances vented into the atmosphere from an ***exhaust system***.

exhaust gases The gaseous contents of a vehicle exhaust, that is, the total content normally understood to include water vapor and ***particulates***.

exhaust gas analyser Instrument for scientifically identifying or analysing constituents of the exhaust gas, used mainly for research or validation.

exhaust gas recirculation Mixing of exhaust gas with intake air to increase the specific heat of the charge and thus reduce the formation of ***oxides of nitrogen***. Also ***EGR***.

exhaust manifold Heat-resistant ducting that connects the exhaust ports of an engine to an exhaust pipe. See ***manifold***.

exhaust pipe Pipe that conveys the exhaust gases away from the engine. See also ***exhaust manifold***; ***exhaust system***; ***manifold***; ***muffler***; ***silencer***; ***tail pipe***. See Figure E.2.

exhaust silencer See ***muffler***; ***silencer***.

exhaust stroke Motion of a piston in an internal combustion engine that expels burned gases from the cylinder.

exhaust system Assembly through which engine exhaust gases pass to atmosphere. The exhaust system may include ***manifold***, ***exhaust pipe***, ***silencers***, ***emission control devices***, ***turbocharger***, devices for the monitoring, control or utilization of exhaust gases, and attachments and mountings of the system. See Figure E.2.

exhaust turbocharging See ***turbocharger***.

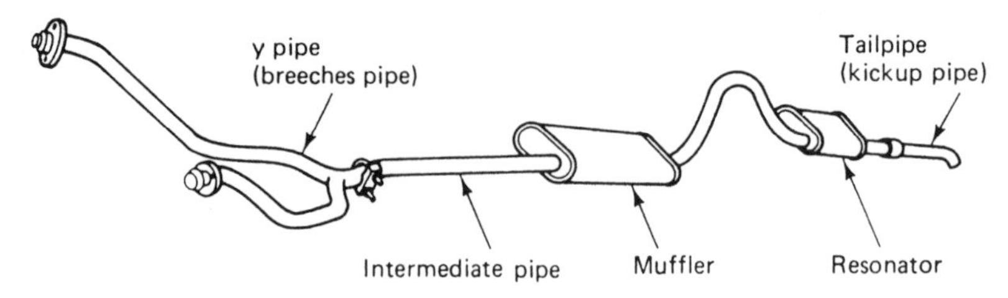

Figure E.2 Exhaust system terminology.

exhaust valve Valve that releases burned gases from a cylinder. See ***overhead valve***, ***side valve***.

exhauster Device for assisting the discharge of bulk cargoes, such as powders, by creating a partial vacuum.

expander Drum brake mechanical or hydraulic mechanism that forces the ***shoes*** apart so that they contact the inner periphery of the ***drum*** and so provide the braking force. See also ***cam-actuated brake***; ***wedge expander***.

expansion tank (1) Tank or container in which engine cooling water boiled off from the main ***radiator*** condenses before returning to the radiator system. (2) Any tank or container that catches overflows resulting from expansion of gases, liquids or vapors.

experimental safety vehicle A prototype or test vehicle built to investigate or assess safety features. Also ***ESV***.

explosion Very rapid combustion, characterized by a sonic wavefront and a sudden loud noise. Damaging to IC engines. See also ***backfire***; ***detonation***; ***knock***; ***pre-ignition***.

extendible See ***extendible trailer***. (Informal)

extendible trailer Commercial vehicle ***trailer*** or ***semi-trailer*** which can be extended to suit the load being carried. Mostly used for conveyance of loads of exceptional length.

extension housing Casing or housing enclosing an extended ***transmission*** mainshaft and sometimes accommodating the ***gearshift lever*** and associated mechanism.

external combustion engine Engine in which the fuel is burned outside rather than within the cylinder, a working fluid such as air or steam being heated by the combustion, for example in the Stirling and Rankine (steam) engines.

extreme pressure Term used to describe high surface-to-surface loads, particularly in lubricated geared systems, such as ***hypoid rear axle*** drives. Also ***EP***.

eye (1) The end of a link of other essentially linear component which is equipped with a flat lug with a hole by which it can be attached. (2) Enclosed looped end of a ***leaf spring*** by which the spring is directly or indirectly attached to the vehicle. Also ***eye end***. See also ***shackle***. (3) Metal former about which the termination of a wire or fibre rope is swaged, to form a secure permanent loop by which a load can be carried.

F

F-head engine (UK: overhead inlet, side exhaust) Engine having overhead intake valves and side exhaust valves, or overhead ***exhaust valves*** and side ***inlet valves***. See Figure F.1.

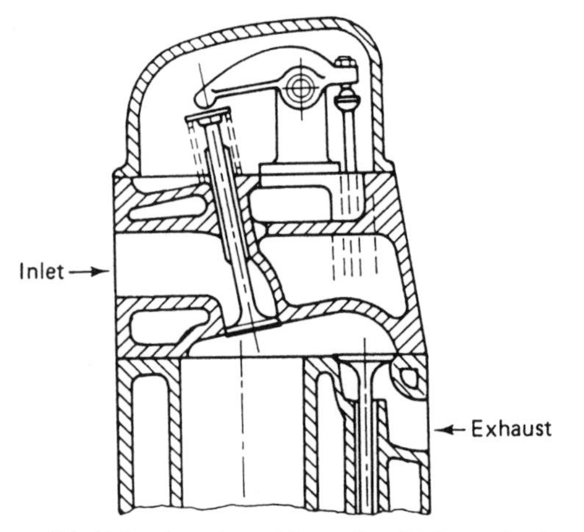

Figure F.1 F-head engine with overhead inlet and side exhaust valves.

fade Reduction of braking effort resulting from overheating or other transient effect.

fail-safe spring brake Commercial or heavy vehicle ***pneumatic brake*** system that automatically applies the brakes under spring load in the event of pneumatic failure.

fairing Any panel that joins other panels with a smooth or fair curve, particularly where this may effect a reduction in drag.

Falex Test Bench test for ***EP lubricant*** properties.

fan Device with rotating blades for moving air. Types found in automotive applications include the conventional multi-blade propeller type, the cylindrical or tangential fan, and the centrifugal or radial fan.

fan belt Endless belt, usually of V or multi-V section, that transmits power from the engine to the ***cooling fan***.

fanfare horn Acoustic warning device operating on electro-pneumatic system, in which an electrically actuated diaphragm excites a tuned column of air, thus giving a clear and penetrating tone. See also ***impact horn***.

fascia Panel or molding immediately below ***windshield*** on vehicle interior, usually facilitating mounting of instruments, air vents and accessories. Not facia.

fast idle The high idle speed of a cold choked engine.

fastback Passenger car with shallow sloping back in which rear screen is mounted.

feather wear Wear of a ***tire tread*** element characterised by thin strips of rubber on one edge of the element.

Federal bumper Bumper specially designed to meet the US Federal Safety Regulations which require the absorption of all the energy of a 5 mph impact.

Federal version In the US, a vehicle that meets Federal emission standards, but not necessarily the more stringent regulations of states such as California.

feed pump Pump that moves a fluid such as a ***fuel*** at a controlled or metered rate.

feed system The pump, piping, valves, and other items that provide a controlled or metered supply of fluid, as from a fuel tank to an engine.

fender (1) **(UK: wing)** Any fixed side-panel of a motor vehicle that partially shrouds a road wheel. See Figure B.4. (2) Deflector plate or structure mounted at the front or rear of a vehicle near ground level.

field of visibility Spread or angle through which an indicating lamp can be seen by an observer. Also ***driver's field of view***.

fifth wheel (1) Coupling table located towards the rear of a truck ***tractive unit***, that bears the weight of the forward end of a ***semi-trailer***, and provides freedom of articulation while acting as a positive towing linkage. See also ***coupling hook***; ***kingpin***; ***wedge-lock***. (2) Towed calibrated wheel used to determine the true speed of a vehicle under test. See Figure F.2.

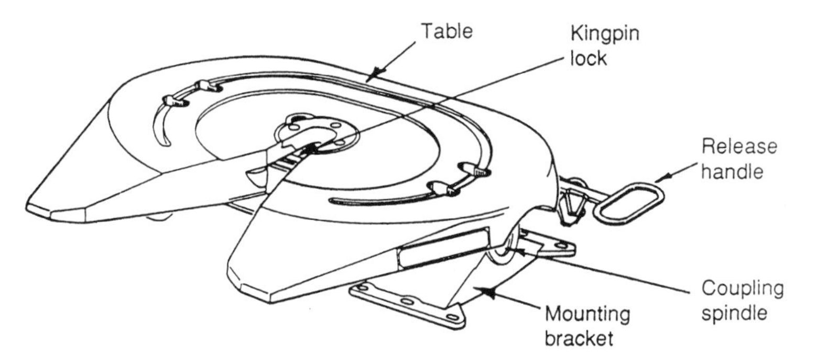

Figure F.2 A fifth wheel coupling.

filler cap Manually removable lid or seal on a filler neck whereby a ***fuel tank***, ***radiator*** or other reservoir is filled.

film strength The measure of a lubricant's ability to remain intact under conditions of ***boundary lubrication***.

filter Porous material, or device containing such material, for removing suspended particulate matter from a fluid, as for example an ***air filter***, ***fuel filter*** or ***oil filter***. Centrifuges and magnetic devices for removing metallic dust are sometimes, though inaccurately, referred to as filters. See also ***cyclone***; ***separator***.

filter element Replaceable porous component of a ***filter***, often of pleated paper, ceramic or fine wire mesh.

fin (1) Thin metal plate protruding from a hot surface to improve the dissipation of heat, as from the cylinder of an air-cooled engine. (2) Vertical aerofoil attached to the rear of a vehicle to improve ***directional stability***. See Figure A.2.

final drive Final geared assembly in a vehicle transmission system, usually the ***differential***.

final drive ratio Speed ratio between the ***propeller shaft*** and the ***driven wheel*** axle shaft.

fineness ratio Ratio of length to thickness, as of a vehicle. See also ***aspect ratio***; ***bluff body***.

finger Lever-type cam follower in an ***overhead camshaft*** engine that transmits the motion of the cam to an intake or exhaust valve.

finning Arrangement of fins on a hot surface such as an engine, radiator or oil cooler. See Figure A.2.

fire appliance A fire-fighting vehicle. Also ***fire engine***. (Informal)

fire ring Annular swelling molded into a cylinder head ***gasket*** to seal against cylinder pressure.

firewall (UK: bulkhead) Transverse panel between engine compartment and passenger compartment intended to inhibit spread of fire to the passenger compartment. See Figure B.4.

firing order The numbered sequence in which the cylinders of a multi-cylinder engine fire.

firing stroke The working stroke of an engine in which the fuel is burned and energy imparted to the piston.

five-mile-an-hour bumper See ***Federal bumper***; ***safety bumper***.

fixed cam brake Drum brake in which the expander mechanism is fixed to the back plate so that the movement of the two shoes is equal. Use of the term is not restricted to brakes with rotating cam mechanisms. See also ***floating cam brake***. See Figure C.1.

fixed choke carburetor Carburetor with a constant size ***venturi***, air flow being controlled by a throttle valve. Also ***variable depression carburetor***; ***fixed venturi carburetor***; ***open choke carburetor***.

fixed control Method of testing the mechanical stiffness of a ***steering*** system in which one element of the steering train is held fixed. See ***fixed steering control***.

fixed head Engine in which the ***cylinder head*** and ***cylinder(s)*** form one inseparable unit.

fixed steering control Method of track testing of vehicle in which the driver holds the ***steering wheel*** in a fixed angular position while the vehicle is subjected to disturbing forces such as lateral wind gusts. See also ***free steering control***.

flag terminal Flat electrical terminal set at right angles to its cable, like a flag, and usually fastened with a bolt or screw. Mainly used for high current applications, such as battery terminals.

flap (1) Horizontal transverse ***aerofoil***, usually with facilities for control or adjustment of angle of incidence, and usually for directing air flow over a vehicle rather than for generating lift. (2) Controllable rear part of an aerofoil for varying the lift or drag of the aerofoil.

flame ionization detector Instrument used in the detection and measurement of ***exhaust gases***, and particularly ***hydrocarbons***.

flame trap Device for preventing the escape of burning gases, as for example from a ***crankcase*** or ***rocker cover***.

flash point The lowest temperature at which the vapors of a flammable liquid product will ignite, though not necessarily remain under continuous combustion, on application of a small flame under standard test conditions.

flasher unit Electromechanical or electronic cyclic switch which causes direction ***indicator (turn signal) lamps*** to flash at appropriate rate. See also ***turn signal lamps***; ***hazard warning lamps***; ***turn signal lamps***.

flat Commercial vehicle or trailer with flat load carrying platform, and usually without side panels or tail board. A ***platform lorry*** or ***truck***. Also ***float***. See Figure T.10.

flat battery Discharged or partially discharged battery incapable of starting a vehicle or providing adequate lighting. (Informal)

flat engine Engine in which the cylinders are disposed in a horizontal plane, and particularly where they are horizontally opposed, as in a flat twin or flat four. See also ***boxer engine***; ***horizontal engine***; ***horizontally opposed***. See Figure F.3.

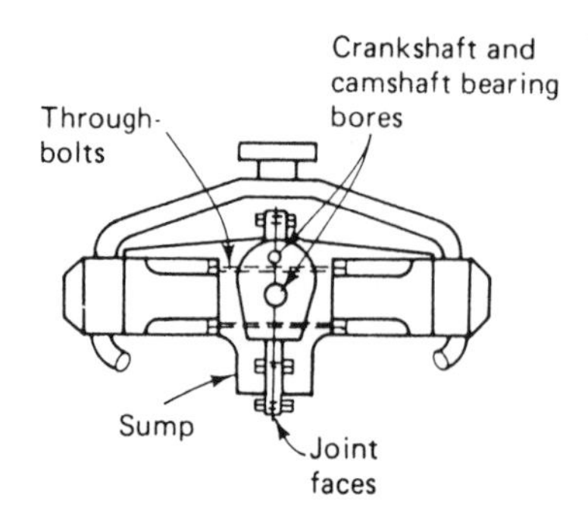

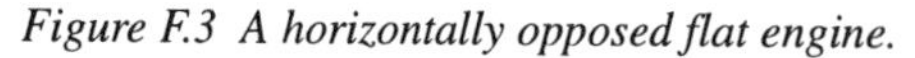

Figure F.3 A horizontally opposed flat engine.

flat four Engine with two pairs of horizontally opposed cylinders. Likewise flat six, flat eight, etc.

flat head (1) Engine combustion chamber, the head of which is flat rather than profiled as for example a wedge or hemisphere. (2) A ***side valve*** engine. (US informal)

flat out Operating at full throttle or at maximum speed. (Informal) ***On the wood***. (US slang)

flat spot (1) A transient reduction in torque of an engine on acceleration, often manifested as a hesitation. (2) A flat patch on a ***tire***.

flatbed truck (UK: platform lorry) Truck with a flat platform to which cargo can be lashed.

flexible Of an engine, exhibiting good torque characteristics throughout its speed range, and particularly the ability to pull at low speed. The informal antonym is ***peaky***.

flexible brake hose High-pressure hose to connect brake pipe line from ***master cylinder*** to ***wheel cylinder*** and allow for wheel movement.

flexible joint Joint or shaft coupling made of flexible material, such as fabric or rubber, fastened between two spiders and capable of transmitting torque through limited angular misalignment of shafts.

flexible rack ***Windscreen (windshield) wiper*** actuating mechanism consisting of a flexible rod usually wound with wire to form a simple rack to engage with the pinion wheels of the wiper drives.

flinger See ***oil flinger***. Also ***slinger***.

flipper Tire ply that wraps around the ***bead bundle*** but does not extend beyond the ***sidewall***. Also ***ply turn-up***. See Figure B.3.

flitch Reinforcement to add strength and stiffness to a frame, for example a chassis member.

float (1) Buoyant part of a fluid metering system, for example ***carburetor***. See also ***float needle***. See Figure C.3. (2) Low platform vehicle or trailer, but particularly for mounting displays in processions, or for conveyance of foodstuffs or drink, as for example a milk float. Also ***flat***. A horsebox. (Aust.) (3) The buoyant member of any fluid level metering or gaging system, as for example a fuel level gage. (4) Tendency of an engine ***poppet valve*** to remain in a constant bouncing state at high speed.

float bowl (UK: float chamber) Main fuel reservoir in a ***carburetor*** which contains the float. See Figure C.3.

float chamber See ***float bowl***.

float needle Needle valve actuated by a float, as in a ***carburetor***. See Figure C.3.

floating caliper Disc brake caliper in which the pinching action of the ***pads*** on the brake disc is achieved by energising one pad only, the ***caliper*** being free to "float" so that the movement of one pad brings both pads into contact with the disc.

floating cam brake Drum brake in which the expander mechanism is not fixed to the back plate, thus enabling it to exert equal loading on the two ***shoes***, though the movement will be unequal. Despite its name, this type of brake does not necessarily use a rotating cam. See also ***fixed cam brake***.

flooding Condition that prevents starting of an engine when more fuel is drawn in than can be ignited.

floor (1) The base panel of a passenger car. Also ***floor panel***. (2) Cargo carrying surface of a truck or van. Also ***loadfloor***. See also ***platform***.

floor panel See ***floor***.

floor shift Gearshift system in which the ***gear selector lever*** is mounted on the vehicle floor or ***transmission tunnel***, rather than on the ***steering column***. ***Stick shift***. (US informal)

flotation Ability of a vehicle to move controllably on mud, or similarly fluid terrain.

fluid clutch See ***fluid flywheel***.

fluid coupling Hydrodynamic coupling by which power can be transmitted, though without the ability to multiply torque as of a torque converter.

fluid flywheel Fluid coupling in which power is transmitted from the driving to the driven rotating elements by hydrodynamic forces on vanes from a circulatory flow of oil in an annular chamber. See Figure F.4.

flutter See ***wheel flutter***.

fly-screen A small windscreen on a ***motorcycle***.

flywheel Massive wheel or disc attached, for example, to the ***crankshaft*** of an engine, to store energy of rotation and smooth the output from the irregular firing of

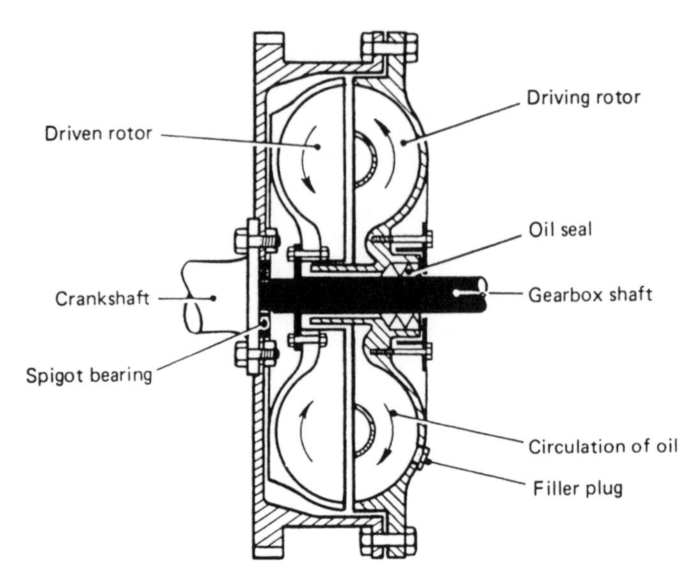

Figure F.4 Sectional view of a fluid flywheel.

the cylinders. In many automotive engines, the flywheel incorporates the ***ring gear*** and acts as one friction face of the ***clutch***. See Figures C.4. and V.1.

flywheel magneto Magneto installed within flywheel, commonly used in small two-stroke engines.

fog lamp Supplementary lamp to provide illumination forward (or rearward) in conditions of poor visibility.

follower Part of a mechanism that is directly driven by a cam, and that imparts motion to the working components of that mechanism. See ***finger***; ***tappet***.

foot brake Brake operated by a pedal.

footprint The shape of the contact interface of a loaded tire with the ground. See also ***contact length***.

footprint aspect ratio Footprint length divided by width, times 100.

footwell Lower interior part of the ***scuttle*** of a vehicle which accommodates the feet of the driver or front passenger, and control pedals. Also ***foot-well***.

force control (1) Mode of testing a steering system in which external forces not resulting from normal steering loads are applied to elements of the steering train. (2) Mode of vehicle control wherein inputs or restraints are applied to the steering system in the form of forces independent of the displacement required.

forced downshift Manual overriding of lower gear selection in an ***automatic transmission***. Also ***kickdown***.

forced lubrication Lubrication by a pressure feed system.

forced ventilation Ventilation by motor assistance, such as an electric fan.

forklift truck Industrial vehicle equipped with two forward-facing arms or prongs which can be mechanically raised and lowered, and on which heavy items can be lifted. See also ***pallet truck***.

forward control (US: cab-over-engine or **COE)** Controlled from the front of the vehicle, particularly of a commercial vehicle in which the cab and driver are located ahead of the engine and front axle. See Figures C.13 and R.3.

forward shoe See ***leading shoe***.

fossil fuel Fuel derived from liquids or gases as the product of decomposition of vegetation or animal remains from an earlier geological period.

Föttinger coupling A fluid ***torque converter*** or hydrodynamic coupling.

fouling Severe contamination, as of a ***spark plug*** with carbon and unburned lubricant.

foundation brake (1) Brake mechanism excluding those parts that rotate with the braked wheel. (2) As (1), but including the ***brake drum***.

four axle rigid Commercial vehicle with load carrying ability and four axles, any or all of which may be driven. See ***rigid truck***. See Figure R.6.

Four Ball test Mechanical test to evaluate wear properties of a ***lubricant***.

Four-square test Test in which the pressure between a pair of gears in contact is used to assess the lubricating qualities of an ***oil*** or ***grease***.

four spring bogie Bogie undercarriage in which each axle is carried on two springs. See also ***four spring non-reactive suspension***; ***four spring reactive suspension***.

four spring non-reactive suspension Tandem axle suspension in which the rear (trailing) ends of the springs are coupled by a ***reaction rod***.

four spring reactive bogie See ***four spring reactive suspension***.

four spring reactive suspension An interactive (reactive) ***tandem axle*** heavy vehicle suspension in which both axles are suspended on ***leaf springs***, the two adjacent springs being linked by a pivoted beam called an ***equalizer beam*** (also called ***balance beam*** and ***walking beam***). See Figure T.1.

four-stroke cycle Thermodynamic cycle of engine operation which requires four strokes of the piston, the strokes usually being designated:

1. induction (inlet, or intake)
2. compression
3. ignition (also called working, power or expansion stroke)
4. exhaust

The majority of vehicle engines operate on the four-stroke cycle. See also ***two-stroke***.

four-wheel drive Transmission system in which engine power is delivered to front and rear wheels of a vehicle. Also ***4WD***.

fraction The product of distillation of a ***crude oil*** or ***petroleum***.

frameless construction See ***integral body construction***; ***unitary construction***.

frazing Removal of unwanted edge or flashing from a forging.

free control Method of mechanically testing a ***steering system*** in which no external forces are applied.

free-field wedge Scientifically designed sound absorbing wedge-form with which an ***anechoic chamber*** is lined.

free-floating pin ***Piston pin*** or ***gudgeon pin*** that is free to rotate in ***piston boss*** and connecting rod ***small end***.

free power turbine Gas turbine in which the turbine that provides the shaft power is not mechanically connected to the compressor and primary turbine stage.

free steering control Method of track testing of vehicle in which the driver releases the ***steering wheel*** when the vehicle is subjected to disturbing forces such as lateral wind gusts.

free travel Of ***brake*** or ***clutch*** mechanism, the amount of pedal movement before the mechanism is actuated. See also ***play***.

freewheel Device that disengages the engine from the drive train on overrun. Obsolete and illegal for motor vehicles in many countries. Occasionally called ***one-way clutch***. See also ***sprag clutch***.

freeze (UK: seize) Sudden adhesive or frictional locking of parts normally in lubricated sliding contact, due to surface welding or clamping, as of a ***piston*** in its bore.

Freon See ***chlorofluorocarbon***.

friction clutch Clutch in which engagement is achieved by friction between rotating surfaces in contact under pressure, as in the conventional ***single plate clutch***, ***multi-plate clutch*** and ***cone clutch***. Progressive engagement and disengagement are achieved by varying the contact pressure, usually by means of a pedal-operated mechanism. See also ***friction plate***; ***sprag clutch***. See Figure C.4.

friction disk damper Damper, mainly used for suspensions, in which angular movement of a knuckle joint of two levers, effectively straddling the suspension spring, is retarded by a friction disk, usually adjustable. (Obsolete) See Figure G.2.

friction drag Aerodynamic drag resulting from the friction between the moving air and the surface of a vehicle in motion.

friction horsepower That part of the total power of combustion of an engine that is spent on overcoming mechanical (or mechanical-and-fluid) friction. See also ***brake power***.

friction lining High-friction wear-resistant material as used in ***clutches*** and ***brakes***. See Figure C.4.

friction plate Clutch disc to which high-friction material is attached. Usually the ***driven plate*** of a ***friction clutch***. See Figure C.4.

front corner marker lamp Lamp set on front corner of a ***trailer*** or ***semi-trailer***, normally visible from forward to 90 degrees outboard.

front-end octane Research Octane quality of the lower boiling-point components of a gasoline. Sometimes used as a specification point to ensure that ***knock*** due to fuel segregation in the inlet manifold under full-throttle acceleration does not occur.

front-end ram Telescopic ram mounted at the front of the ***tipper body***, so that the load is discharged at the rear of the vehicle.

front-wheel drive Transmission system in which the engine power is delivered to the front wheels of a vehicle.

fuel Combustible form of energy for an engine, usually in liquid or gaseous form.

fuel dragster Drag-racing machine using non-gasoline based fuel.

fuel cell Device that converts gaseous or liquid fuels into electrical energy through electrochemical rather than mechanical agency.

fuel consumption Rate of consumption of fuel by an engine, expressed in units such as miles per gallon or litres per kilometer. See also ***specific fuel consumption***.

fuel distributor Device which meters and directs fuel to the ***injectors*** of a fuel-injection engine. See also ***distributor pump***.

fuel economy oil Engine lubricant which enables an engine to consume less fuel than a standard ***lubricant***, usually by reduction of ***viscosity*** and increase of ***additive*** content.

fuel gage (UK: fuel gauge) Instrument for indicating the amount of fuel in a ***fuel tank***.

fuel injection Injection of fuel under pressure, into the intake tract, directly into the cylinder, or indirectly into a cylinder ***pre-chamber***. See also ***direct injection***; ***indirect injection***.

fuel injection pump Mechanical pump that provides, and in many instances meters, fuel at pressure to ***fuel injectors***. Also ***supply pump***. See also ***distributor pump***; ***in-line pump***.

fuel injector Device whereby fuel is injected in metered quantities into an engine. See Figure I.3. See also ***fuel injection pump***.

fuel line The route of the fuel supply from ***fuel tank*** to ***engine***.

fuel pressure regulator Pressure-actuated diaphragm valve that maintains the pressure in a fuel system to a pre-set value above ***manifold pressure***, particularly in a ***fuel injection*** system.

fuel pump (1) Mechanical or electrical pump that draws fuel from a tank to provide the fuel supply for ***carburetor*** or ***fuel injection system***. (2) The high-pressure ***fuel injection pump*** of a ***diesel engine***.

fuel sac Cavity in a diesel ***fuel injector*** between the needle valve seat and the spray holes. See Figure I.3.

fuel system Combination of ***fuel tank***, fuel lines, pump, filter and vapor return lines, carburetor or injector components, and all fuel vents and evaporative emission control devices or systems.

fuel tank Reservoir or container from which an engine draws its fuel.

fulcrum pin See ***kingpin***; ***swivel pin***.

fulcrum ring Circular fulcrum on which the ***diaphram spring*** pivots in a ***diaphragm clutch***. Also ***diaphragm spring ring***. See Figure D.3.

full-flow filter Filter through which the total flow of fluid passes, as opposed to the ***by-pass filter***, which filters only part of the flow.

full fluid film lubrication Lubrication in which both contingent surfaces are separated by a continuous film of lubricant. See also ***boundary lubrication***.

full-load enrichment device ***Carburetor*** system whereby fuel is supplied directly from ***float chamber*** to ***barrel*** at high load demand.

full-time drive ***Four-wheel drive*** system without facility for disengaging one axle, so that all wheels drive all the time. (US informal) See also ***part-time case***.

full trailer An independent steerable trailer with at least two axles. A trailer so constructed that no part of its weight, except the towing device, rests upon the towing vehicle. See also ***semi-trailer***.

fully floating axle Rear axle in which the axle ***half shafts*** serve only to transmit torque to the wheel, the total vehicle weight and cornering loads being transferred directly from the wheel bearings to the axle casing. Also ***full floating axle***. See Figure F.5.

fuse Replaceable device which opens an electrical circuit with irreversible action when the current exceeds a predetermined value. See also ***circuit breaker***.

fuse box Covered panel or container for vehicle electric circuit fuses.

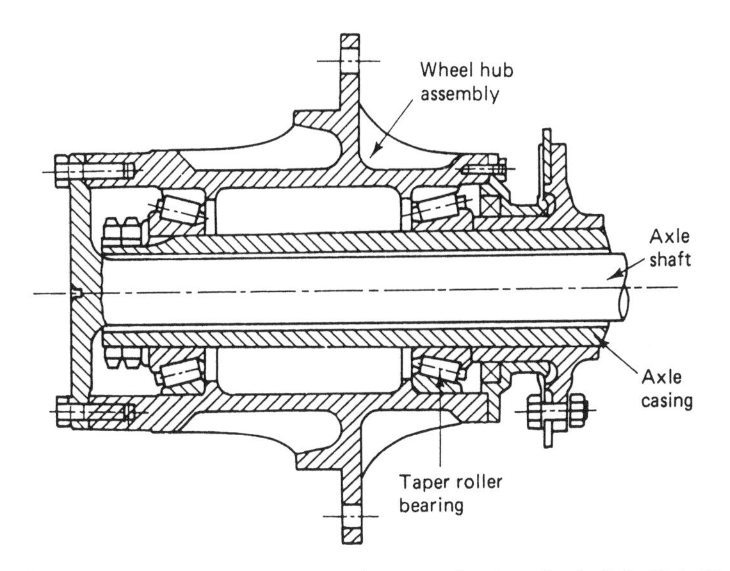

Figure F.5 Sectional view of a fully floating axle: the axle shaft (half shaft) and wheel hub. The axle casing remains stationary.

fuse rating Indicated current capacity of a ***fuse***, such that the fuse will fail at a specified percentage of rated current after a specified time.

G

G-string See ***start-up groan***.

gaiter See ***boot***.

gangway Passage between rows of seats on a ***bus*** or ***coach***.

gas (UK: petrol) Gasoline motor fuel. (US informal) Also ***motor spirit*** in mainly British official usage.

gas chromatography System of analysis of gases or vaporised liquids, such as ***exhaust gases***, using the gas chromatograph which separates and identifies individual components of a mixture according to their tendency to be adsorbed, in a column or a capillary tube.

gas generator (1) Device in which combustible gas is produced by burning a solid fuel. In times of acute shortage of liquid fuel, vehicles have been converted to run on gas from a "producer," burning a solid fuel such as coal or anthracite. This system was prevalent in Europe during the Second World War. (2) Gas-producing unit for starting a power gas turbine, often in the form of a compact, simple auxiliary gas turbine. (3) Simple gas turbine for providing gas for a mechanically independent power turbine driving a shaft. Also called ***gasifier***. (4) A two-stroke pressure-charged engine producing pressurized hot gas for expansion through a gas turbine.

gas guzzler Car with high fuel consumption, and particularly one dating from before the period of energy conservation. (US slang)

gas oil Heavy petroleum distillate used as a diesel fuel and as a blending stock for fuel oil, often graded as light gas oil (LGO), light catalytic gas oil (LGCO) and heavy gas oil (HGO). Term mainly used in refinery parlance. Believed to be so called because it was used in vaporised form to supplement town's gas.

gas spring Spring, particularly a suspension spring, using gas under pressure as a spring medium. See also ***air bellows***.

gas tank (UK: petrol tank) (1) Vehicle-installed gasoline fuel tank. (US) (2) A vehicle-installed tank for liquefied petroleum gas or other gaseous fuel. (UK)

gas truck (UK: petrol tanker) Gasoline delivery tanker. (US informal)

gas turbine Internal combustion engine in which the energy released by burned gas drives a turbine. The turbine is usually directly coupled to a compressor which increases the pressure of the air entering the combustion chambers, and in so doing increases the thermal efficiency. When used in aircraft propulsion the gas turbine is often informally called a jet engine. This term is valid only when the jet thrust is the main agency of propulsion. In automotive applications the gas turbine drives the roadwheels through a transmission.

gasifier See ***gas generator***.

gasket A static seal used to contain pressure and prevent leakage. In automotive terminology a gasket is usually a flat, compressible seal, as for example a cylinder head gasket.

gasohol Automotive fuel normally consisting of a nine part to one blend of ***gasoline*** and ***alcohol***.

gasoline (UK: petrol) Light hydrocarbon fuel used in spark ignition (SI) engines. Also ***gasolene*** (UK archaic); ***gas*** (US informal).

gasoline engine (UK: petrol engine) Internal combustion engine in which gasoline fuel is vaporised and mixed with air before compression and initiation of combustion by a spark. A spark ignition engine using gasoline or petrol as a fuel.

gate Mechanical constraint to gear lever movement to ensure accurate engagement and prevent possibility of partially engaging two gears at once.

gate change Manual or semi-automatic gear change constrained by a gate.

gather See ***toe-in***.

gauze carburetor Early form of ***carburetor***, in which induction air gathered fuel vapor from a gauze over which gasoline flowed. See also ***wick carburetor***.

GCW See ***gross combination weight***.

gear and pinion steering See ***worm and sector steering gear***.

gear cluster A countershaft or layshaft gear assembly. Also occasionally ***cluster gear***.

gear lever See ***gearshift lever***. (UK informal)

gear ratio Ratio of angular velocities of pairs of meshing gears.

gear step Difference in ratio between two adjacent gears.

gear train Series of meshing gears designed to achieve a given overall gear ratio.

gear wheel Toothed wheel used to transmit power without slip.

geared speed (1) Theoretical speed of a vehicle derived from engine speed, overall geared ratio to driven wheels and diameter of driven wheels. (2) Maximum speed that a vehicle can achieve in each gear. (Non-preferred definition) See also ***powered speed***.

gearbox (US: gearcase, transmission) Encased assembly of gears, but particularly a ***manual shift*** or ***pre-selector unit***. UK terminology favors use of term ***trans-***

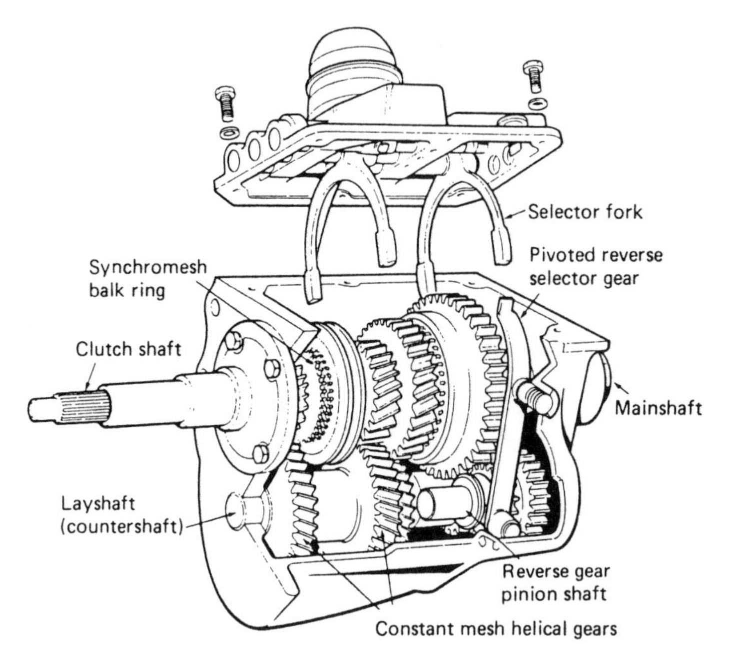

Figure G.1 A Ford four-speed synchromesh gearbox.

mission in context of ***automatic transmission***, but otherwise uses the term transmission to refer to the drive train system from clutch to final drive. See Figure G.1.

gearcase See ***gearbox***.

gearset Assembly of ***crown wheel***, ***pinion*** and ***differential gears*** of a transmission.

gearshift lever Lever for manually changing gear in a change-speed transmission. Also ***gear lever***. See also ***column change***; ***floor shift***.

gearstick Gearshift lever. (Informal)

Gemmer steering gear Proprietary steering gear of the ***hourglass*** or waisted worm and roller type.

generator Rotating electrical machine for producing current at a nominal voltage, such as an ***alternator*** or a ***dynamo***.

geometric displacement Calculated displacement or capacity of a ***Wankel engine***. See ***cell swept volume***.

geometry In context particularly of ***steering*** and ***suspension***, the angular and linear relationships between the principal items and changes thereof in operation.

gills Flaps or louvers, often adjustable, to facilitate ventilation. (Informal)

girder fork Form of single wheel suspension in which the wheel is straddled between a pair of pivoted struts, one carrying the wheel hub and the other attached to the frame or chassis of the vehicle, relative motion being reacted by a spring. A feature of earlier motorcycles. See Figure G.2.

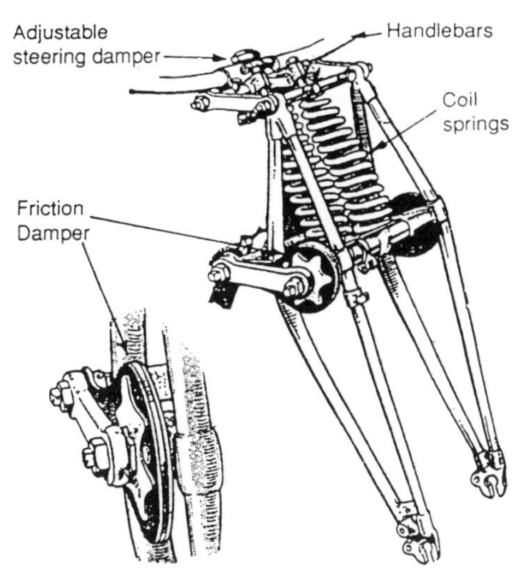

Figure G.2 Webb motorcycle girder forks of 1924, showing parallelogram action, and with friction dampers inset.

girder spindle Pivot spindle of a ***girder fork*** front suspension.

gladhand Multiple connector for pneumatic hoses, as for example between a ***tractor*** and ***trailer***. (US informal) See also ***suzie***.

glove box Recess in or below the ***fascia*** for holding gloves or other small items.

glow plug (1) Electrically heated plug fitted to the cylinder head of a diesel engine. An element provides heat within the combustion chamber and aids starting in cold conditions. (2) A similarly installed plug that aids starting and becomes incandescent during the operation of an engine to initiate or aid combustion. See also ***hot bulb ignition***; ***semi-diesel***.

gooseneck (1) The pivoted coupling arm of a vehicle trailer. (2) Stepped forward part of a ***step-frame*** semi-trailer.

Gough-Joule Effect The property of rubber when in tension to contract when it is heated, which is the opposite to most engineering materials.

governor A device which limits the maximum speed of an engine. Mainly found on ***diesel engines*** to prevent mechanical damage or operation at speeds that would give rise to unacceptable smoke emission.

governor valve Valve that controls air output from the compressor of an air brake system. See also ***unloader valve***.

grab Sudden unduly high output from the ***brakes***.

grabber See ***seat belt grabber***.

gradeability Measure of the ability of a vehicle to ascend an incline.

graphite Naturally occurring form of carbon, used, because of its softness, as a solid lubricant or grease additive. (Mainly UK usage) US usage of the term tends to embrace various forms of carbon, as for instance as a reinforcing medium for plastics, that would not be classed as graphite in UK. Synthetic production of graphite is possible.

Gran Turismo A powerful touring car, often abbreviated to GT.

grease Solid form of ***lubricant*** which remains in place when undisturbed, but flows when subjected to motion, as in a bearing. See also ***lithium grease***; ***rheology***.

greenhouse effect Atmospheric condition arising from the entrapment of radiation as by the roof of a greenhouse, leading to a rise in ambient temperature. See also ***greenhouse gases***.

greenhouse gases Infrared absorbing gases that contribute to the ***greenhouse effect***, principally ***carbon dioxide*** and ***chlorofluorocarbons***.

grille Decorative and protective grid at front of vehicle through which air passes to enter the engine compartment. Sometimes ***grill*** or ***radiator grille***.

groove Narrow void in tire tread pattern. See also ***sipe***.

gross axle weight Specified maximum carrying capacity of an axle, measured at tire-road interface. Also ***gross axle weight rating*** (GAWR) in US.

gross combination weight Total weight of towing vehicle and trailer (or semi-trailer) with equipment, fuel, driver and passengers, payload, and trailer. Also ***GCW***.

gross contact area Total area of tire footprint or ***contact patch***, including area of ***grooves*** or ***voids***.

gross power The measured power output of an engine operating without power absorbing ancillaries such as electric ***generators***, pumps and ***silencers***. The power of a basic engine. See also ***installed power***.

gross train weight Combined weight of towing vehicle with ***drawbar trailer*** and load. Also ***GTW*** or ***gtw***.

gross vehicle weight (1) Maximum legal weight at which a vehicle can be operated. Curb weight plus payload. (2) Combined weight of a truck and its load. Also ***GVW***.

ground (UK: earth) Return electrical circuit, or its attachment point as to chassis or battery.

ground clearance Vertical distance between level ground and lowest fixed item on the vehicle.

ground effect Aerodynamic effect of proximity of ground to vehicle.

ground effect vehicle Vehicle using the venturi effect between the underbody and ground to aerodynamic effect.

ground line angles The inclined forward, rear and inter-axle ground angles, usually referred to the ***ground plane***, that describe the steepest ramp that the vehicle can approach or leave, or the steepest arched ramp over which the vehicle can pass. See also ***approach angle***; ***departure angle***; ***interference angle***.

ground plane Horizontal reference plane on which a vehicle stands. See also ***road plane***.

grouser Transverse metal tread on track shoe of a track-laying vehicle to increase traction. See Figure C.12.

growl Low-frequency tire ***tread noise***, similar to that produced by traversing a metal grating.

grown tire Tire which has expanded as a result of use.

GTW See ***gross train weight***.

gudgeon pin (US: wrist pin) Bearing pin that connects a ***connecting rod*** to a ***piston***.

gulp valve Valve to introduce extra air to induction tract on acceleration and so prevent over-richness of fuel-air mixture.

gum Adhesive product of poor combustion deposited in the cylinder or exhaust tract of an engine.

Gurney flap Aerodynamic ***spoiler*** running the spanwise along the trailing edge of the ***wing*** of a competition car, usually taking the form of an extension at right angles to the chord of the trailing edge, to create low pressure at trailing edge. After Dan Gurney, American racing driver. See Figure G.3.

guttering Lipped edge at side of roof panel, by which rain water is channelled away.

GVW See ***gross vehicle weight***.

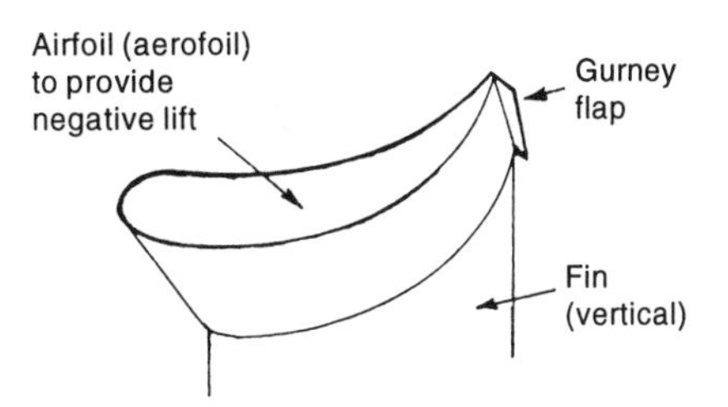

Figure G.3 View on center section of a race car wing, showing fin and Gurney flap.

H

H engine Engine with two sets of opposed cylinder axes geared to a central single drive shaft.

H-pattern shift Gearshift or gearchange movement which follows the pattern of a letter H, usually with first gear at top left, second bottom left, then to third at top right and fourth at bottom right. See also ***gate***; ***gate change***.

H-point Pivot point of the torso and thigh on two- and three-dimensional ***anthropometric dummy*** devices or manikins used in defining and measuring vehicle seating accommodation.

hairpin valve spring Valve spring formed from wire or metal strip bent to form two levers emanating from a half-loop or coil.

half cab Vehicle ***cab*** located on one side of the vehicle's centerline, providing accommodation for the driver only.

half elliptic spring See ***semi-elliptic spring***.

half shaft Shaft by which power is transmitted from final drive to one driven wheel or pair of twin wheels. Also ***half axle***.

half-track Vehicle, particularly a military vehicle, with traction provided by a powered chain track but steered by conventional wheels. Also ***semi-tracked vehicle***. See also ***track laying vehicle***.

halogen Element of the chemical family chlorine, bromine, fluorine and iodine.

hand lay-up Manual system of producing larger GRP components and panels, by applying by brush or spray the resin to the glass fiber reinforcement in an open (one-sided) mold. Labor intensive, and only one surface is smooth finished.

hand-reach envelope Practical extent or reach of a seated constrained driver or passenger.

hand start (1) To start an engine by manually cranking, or by manual operation of an automatic cranking mechanism such as a ***recoil-starter***. (2) Mechanism by which an engine can be manually started.

handbrake Brake operated by a hand lever. A ***parking brake***.

handed lamp Lamp with different inboard and outboard characteristics.

hard plug A cold grade of ***spark plug***.

hard top (1) Sports car with a fixed or rigid roof, as an alternative to a similar ***soft top*** model. (2) Passenger car with fixed rigid roof. A conventional saloon or sedan car. (3) The roof assembly, particularly where it is made as a removable or accessory item to an otherwise open car. See also ***soft top***.

harmonic balancer Rotating or oscillating counterbalance that counteracts the out-of-balance forces and/or couples in a reciprocating engine, or smooths the torsional fluctuations in an engine crankshaft. Numerous types have been devised, such as the Lanchester anti-vibrator, after inventor F.W. Lanchester. See also ***vibration damper***.

harmonic damper See ***harmonic balancer***.

harmonic induction engine Induction system in which the length of the inlet tract is chosen to improve volumetric efficiency over a narrow speed band. Also ***tuned induction***. See also ***ram air***.

hatchback Passenger car with hinged lifting back access door or ***tailgate*** encompassing the backlight or rear screen.

Hayes transmission Infinitely variable transmission of the ***toric*** type in which the ratio change is brought about by caged wheels running within toroidal tracks or races and so pivoted that the ratio of rolling diameters between the driving and driven races can be varied. Also called ***Austin Hayes gearbox***. The transmission replaced the ***gearbox***, but not the ***clutch***, in certain vehicles in the 1930s. The principle is similar to that of the ***Perbury gearbox***.

hazard warning lamps Flashing lamps, mounted one on each corner of a vehicle, to indicate presence of vehicle, particularly in event of breakdown or accident.

hazchem Code of abbreviations for hazardous chemicals displayed on carrying vehicle, principally for information to fire and ambulance services in event of accident or spillage.

hazing Reduction of transparency of ***windshield*** or windscreen through ineffective operation of a wiper blade.

head See ***cylinder head***.

head board Vertical barrier attached to the forward end of a ***platform*** of a truck. See also ***headerboards***.

head gear Primary reduction gear pair providing a speed reduction of the ***layshaft*** or ***countershaft*** in a heavy vehicle transmission.

head lining See ***headlining***.

head pipe Section of an exhaust system immediately following the exhaust manifold. In a vehicle equipped with ***catalytic converter***, the pipe joining manifold to converter. Also ***downpipe***. (Obsolescent) See also ***tail pipe***.

head race The bearing, usually an adjustable ball bearing, of a motorcycle ***steering head***.

head restraint Cushioned headrest or pad firmly supported behind an occupant's head to minimize ***whiplash*** injury in impact.

head-up display Instrument display reflected in ***windshield*** (***windscreen)***, to give driver view of essential instruments without deflecting gaze to instrument panel.

header tank Vessel that contains a liquid at a higher level than the main tank, so that the level or static head of pressure can be maintained.

headerboards Vertical barrier, sometimes of stackable and movable boards, attached to the forward end of a trailer. See also ***head boards***. (Also UK usage)

heading angle Angle by which the longitudinal axis of a moving vehicle deviates from its true direction of motion. ***Yaw angle***.

headlamp Lamp to provide upper (main) or lower (dipped) beam illumination ahead of the vehicle. See also ***dipped beam***; ***driving lamp***; ***main*** or ***upper beam***.

headlamp beam switch Driver-controlled device for selecting upper (main) or lower (dipped) headlamp beam circuit. See ***dipper switch***; ***semi-automatic beam switch***.

headlight The beam of light from a ***headlamp***. Not the lamp itself.

headlining (US: roof lining) Fabric roof lining or ceiling of a vehicle body. Also ***head lining***.

headrest Upward, usually adjustable, extension of a seat back, for supporting the head of a vehicle occupant. See also ***head restraint***.

heat aging (1) The effect of long-term exposure to heat on the mechanical and physical properties of a material. (2) The testing of a material to determine the effect of long-term exposure to heat.

heat control valve Valve that regulates the flow of exhaust gas so that some of its heat content is passed to the intake ***manifold***, thereby helping to vaporize the fuel mixture on starting or for operation in conditions of extreme cold.

heat dam (UK: slotted piston) Annular slot or insert in piston to minimize flow of heat from ***crown*** to other parts of piston.

heat engine Engine deriving energy from the heat of combustion of a fuel, whether burned internally or externally. See also ***external combustion engine***; ***internal combustion engine***.

heat fade See ***fade***.

heat range Range of temperature for optimum operation of a ***spark plug***. See ***heat range index***. See also ***cold plug***; ***hot plug***.

heat range index Standard for the designation of the range of operating temperature for ***spark plugs***.

heat transfer fluid Liquid or vapor of high specific heat for taking heat from one part of a system to another.

heated intake System whereby the induced air or air/fuel mixture in an engine is heated to reduce emissions on starting or to facilitate operation in conditions of extreme cold. See also ***hot spot***.

heavy duty oil (1) Crankcase lubricant with ***detergent additives*** for larger engines. (2) High detergency oil for specific types of commercial ***diesel engine***.

heavy goods vehicle Vehicle intended for ***heavy haulage*** or conveyance of goods and in most countries designated by exceeding a certain unladen weight and requiring of the driver special training and license. Also ***HGV***.

heavy haulage Legal (and often supervised) carriage of loads heavier than normally permitted by commercial vehicle legislation.

heavy locomotive A road ***locomotive*** with an unladen weight exceeding a stipulated figure, nominally 11,690 kg. (Mainly UK usage)

heelboard Lower part of front ***bulkhead***, ***firewall*** or ***scuttle***, particularly where installed as a separate structural item.

helmet connector Connecting cap for a ***battery*** with tapered terminals.

Helmholz resonator Acoustic device, shaped like a jug or bottle, which is resonant at pre-determined frequencies. See also ***quarter-wavelength resonator***.

helper spring Additional spring on a suspension system that operates only at large deflection of the main spring. See also ***chassis stop***.

hemi ***Hemispherical head*** of an engine. (Slang)

hemi-anechoic chamber Anechoic chamber with a reflective floor. Also ***semi-anechoic chamber***.

hemispherical head Engine cylinder head of true or flattened hemispherical form. See also ***cross flow***; ***cross scavenging***.

HGV See ***heavy goods vehicle***.

high beam Headlamp ***main beam***.

high lift cam Special cam profile on engine camshaft to increase ***valve lift***, usually for sporting or competition use.

highway cycle Any standard vehicle test cycle that simulates driving on the open road and predominantly in higher gears. See also ***EPA Highway Cycle***.

hitch (1) Articulating coupling whereby a trailer or agricultural implement is towed or drawn. (2) A simple ball and socket coupling. See also ***drawbar***.

holding valve In a hydraulic circuit, a valve that prevents movement or slippage of a hydraulically supported item, such as an excavating bucket, which is to be temporarily held in one position.

homofocal headlamp Headlamp with segmented or stepped reflector to give different focal lengths, usually used with a two-filament bulb. Use of this term is misleading. Also ***homofocular headlamp***. See also ***bifocal reflector***; ***stepped reflector***.

hood (1) **(UK: bonnet)** Hinged or removable body panel by which access is gained to the engine compartment of a vehicle. See also ***underhood***. (2) **(US: soft-top)** Folding fabric top of a ***convertible***.

hood scoop Duct, usually for the admission of additional cooling air, into the engine bay of a vehicle by way of the ***hood*** (US) or ***bonnet*** (UK). (US informal)

Hooke's joint Universal joint in which the shaft ends are connected by yokes disposed at right angles and communicating torque by way of a cruciform bearing mounting. Also ***Cardan joint***. Also the linking of two such joints to eliminate non-uniform rotational velocity. See Figure H.1.

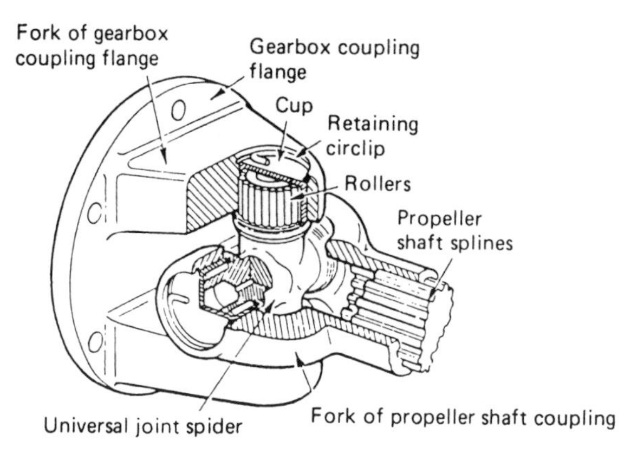

Figure H.1 Flange-mounted Hooke's joint on a drive shaft (propeller shaft).

Hooper engine A variety of ***stepped piston engine***. See Figure S.9.

hop See ***wheel hop***.

hop up (UK: soup up, tweak) To tune or modify an engine for performance. (US slang)

horizontal engine (1) Engine in which the cylinder axes are in the horizontal plane, as of a "conventional" in-line four turned through 90 degrees. (2) A horizontally opposed engine. A ***boxer engine***. (Informal)

horizontally opposed Of an engine, having the ***cylinders*** set in a horizontal plane at either side of the ***crankshaft***. See ***flat engine***; ***horizontal engine***.

horn Audible warning device, usually electrically operated.

horse-box Van or trailer for conveyance of a horse or horses. (Mainly UK usage) A ***horse van*** body (US); ***float*** (Aust.)

horse van body Body type for conveyance of a horse or horses. (Mainly US usage) Also **horse-box**. (UK)

horsepower The customary non-metric unit of power, equivalent to 0.7457 kilowatt, defined by a rate of working of 33,000 foot-pounds per minute. See also ***brake power***; ***indicated power***.

horseshoe vortex Aerodynamic vortex occurring between the vortices of a vortex pair in the wake of a moving vehicle and comprising the transverse and trailing vortices.

hose Laminated, or reinforced, flexible pipe for transport of liquids or gases usually at elevated temperature or pressure, and often with an external protective cover which may or may not be bonded. See also ***suzie***; ***tube***.

hot bulb ignition Ignition system employing a hot or incandescent element, externally heated for starting and in some cases receiving external heat to sustain operation. See also ***semi-diesel***.

hot plug (1) Spark plug that operates at a high temperature in relation to the combustion temperature, thus minimizing the risk of plug fouling in a low compression engine. (2) Thermally insulated lower half of a combustion chamber usually having a tangential throat to generate swirl within the combustion chamber to aid combustion.

hot press molding System of making GRP components from resin and glass fiber preforms, which are formed in heated steel molds brought together under pressure, usually of a hydraulic press.

hot soak losses Fuel vapors emitted during a specified period beginning immediately after the engine is turned off.

hot spot (1) Point of contact between intake and exhaust manifolds to transfer heat to the fuel mixture and thereby promote vaporisation. See also ***heated intake***. (2) Any overheated point on an item.

hot-rod Production car individually modified to give an outward appearance of opulent power, either for road use or for racing.

hot-shift PTO Power take-off device which can be remotely engaged by hydraulic, pneumatic or other powered means. A ***power shift PTO***.

hot-start enrichment Enrichment of fuel mixture for starting a hot engine, usually applicable only to gasoline engines with electronic fuel injection.

hot sticking Adhesion of piston ring to groove caused by formation of deposits at higher temperatures. See also ***cold sticking***.

hot testing Testing an engine in laboratory conditions with the engine running, and with exhaust, cooling and fuel systems connected. See also ***cold testing***.

Hotchkiss drive Rear-wheel-drive transmission system in which a live beam axle and differential assembly is coupled by a universal joint to a propeller shaft, which is in turn coupled by a universal joint to a gearbox. The axle is normally mounted on ***semi-elliptical*** leaf springs. See Figure H.2.

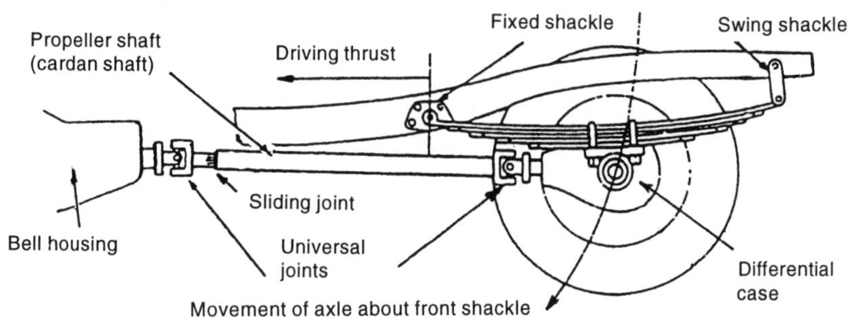

Figure H.2 A simple Hotchkiss drive arrangement.

hourglass worm steering Steering gear in which a waisted worm gear attached to the ***steering column*** imparts angular motion to a toothed sector or roller. See also ***Gemmer steering gear***; ***Marles steering gear***.

hub The center assembly of a wheel containing the wheel bearings.

hub cap Cap fitted to the outer end of a hub to protect bearings. See also ***Rudge cap***.

hub plate Splined central element of a clutch ***driven plate***.

hub reduction Reduction gearing, usually epicyclic, located within a wheel hub, to provide an additional set of ratios in a heavy vehicle, and to reduce torsional loading of half shafts and differential.

huffer A ***supercharger***, particularly in competition context. (US slang) A ***blower***.

hunting Variation of speed of an engine about a mean when governed or at a constant fuel delivery setting.

hush kit Post-production kit of noise-suppressing materials and components.

hybrid engine Engine combining two principal modes of operation, such as that of diesel engine and spark ignition engine, or an internal combustion engine operating with an electromechanical drive.

hybrid vehicle Vehicle employing two distinct but interdependent forms of propulsion, such as an electric motor and an internal combustion engine, or electric motor with battery and fuel cells for energy storage.

Hydragas suspension Proprietary suspension system in which a gas spring is actuated through a diaphragm by fluid under pressure, fore and aft springs on each side of the vehicle being hydraulically linked to share load and minimize pitch.

hydraulic brake Brake actuated by hydraulic pressure.

hydraulic clutch See ***fluid flywheel***.

hydraulic damping Damping by the viscous flow of a fluid through a constricted orifice. See also ***shock absorber***.

hydraulic head assembly The pumping, metering and distributing elements of a ***distributor fuel pump***. See also ***distributor pump***.

hydraulic lifter Small hydraulic actuator which operates the intake and exhaust valves of an engine, either directly or through a mechanical linkage. See also ***hydraulic tappet***.

hydraulic retarder Transmission mounted device using fluid friction to retard the speed of a vehicle or assist the ***service brakes***.

hydraulic tappet Self-adjusting oil-filled tappet that compensates for wear in the ***valve train***. See Figure H.3.

hydraulic transmission Transmission employing a hydraulic ***torque converter***.

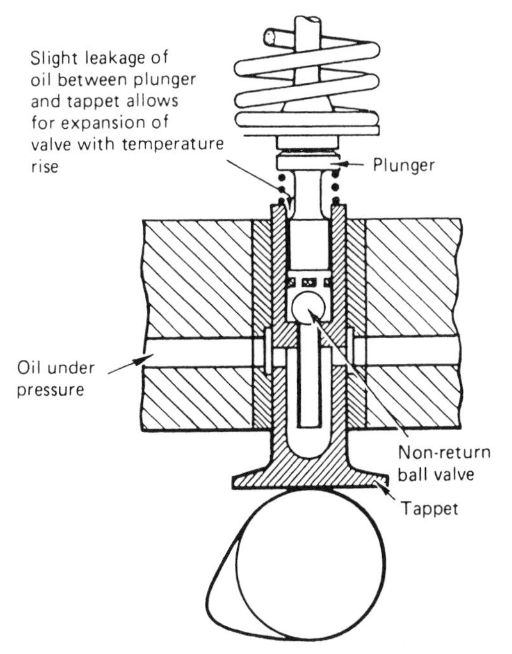

Figure H.3 Hydraulic tappet for an overhead valve engine.

hydro-mount Viscous fluid anti-vibration mounting, particularly for engine mounting.

hydrocarbon Organic compound consisting only or principally of carbon and hydrogen. The main constituent of liquid and gaseous fuels. See also ***unburned hydrocarbons***.

hydrocarbon emissions Unburned or partially burned hydrocarbon fuels exhausted to atmosphere from an engine. Often contracted to HC. One of the principal automotive atmospheric pollutants. See ***THC***.

hydrocarbon trap Device for preventing loss to atmosphere of exhaust unburned ***hydrocarbons***. Sometimes used in conjunction with a ***catalytic converter***.

hydrodynamic drive See ***hydrokinetic transmission***.

hydrodynamic lubrication Lubrication in which the oil film is generated by the relative movement of surfaces and the existence of an oil wedge.

hydrofining Hydrogen refining process for improving products such as lubricating oil.

hydrogen An elementary gas, and component of water and hydrocarbons, with potential as an ***alternative fuel*** for internal combustion engines.

hydrokinetic transmission Power transmission system, such as a ***torque converter***, in which power is transmitted primarily by the motion of a fluid in an enclosed recirculatory path, rather than by static pressure. Also ***hydrodynamic drive***.

Hydrolastic suspension Proprietary interactive suspension system in which loads are reacted by deformation of a rubber diaphragm spring within a sealed fluid-filled chamber, the consequent change of volume of the fluid filled cavity being communicated to the suspension of the second wheel on the same side.

hydroplaning (UK: aquaplaning) Effect whereby a vehicle tire rides up on a thin surface of water and in so doing loses contact with the road surface resulting in sudden loss of ***traction*** and control.

hydropneumatic suspension Suspension employing hydraulically loaded gas springs, often with hydraulic coupling between axles.

hydrostatic transmission Drive by means of hydraulic motors, particularly where the drive is to each wheel of an off-highway vehicle.

hypoid axle Driving axle incorporating a hypoid differential gear.

hypoid gear Bevel gear with the axes of the driving and driven shafts at right angles, but not in the same plane, giving some sliding action between teeth. Widely used in ***differentials***. See Figure H.4.

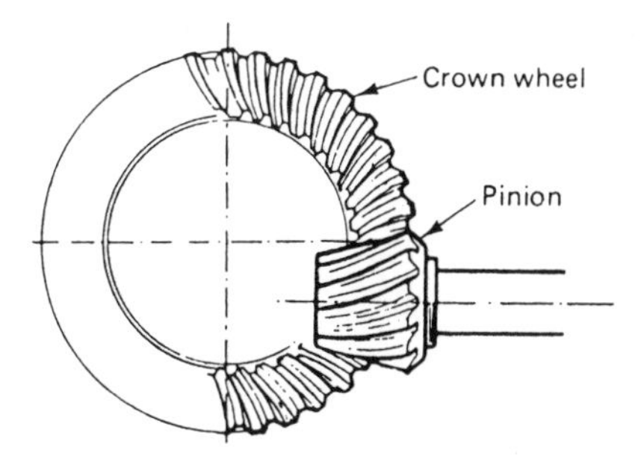

Figure H.4 Hypoid crown wheel and pinion, as from a back axle final drive.

I

I-head engine (UK: overhead valve engine) Engine with ***intake valve*** and ***exhaust valve*** in ***cylinder head.***

IAC See ***idle air control.***

IC engine See ***internal combustion engine.***

identification lamps Lamps used in clusters at front or rear top to identify specific classes of large vehicles.

idle air control Valve for controlling air flow in induction system when engine is idling. Also ***IAC.***

idle speed Rotational speed of an engine on no-load and minimum throttle setting. Also ***tick over.*** (Informal)

idle system Arrangement of jet and tubes in ***carburetor*** to enrich the fuel supply when engine is idling.

idler (1) Any gear between a driving and driven gear in a gear train, the shaft of which serves only to bear and locate the gear, and is therefore "idle." The idler gear may serve to reverse the sense of rotation of the driven wheel, so that it rotates in the same direction as the driving wheel. (2) An intermediate or tensioning pulley in a belt drive. (3) The non-powered wheel that guides and supports the track of a ***track-laying*** vehicle.

idler arm (1) Slave drop arm actuated by a tie rod or center track rod from the steering box drop arm. The idler arm transmits steering action via a further ***tie rod*** or ***track rod*** to the steered wheel furthest from the ***steering box.*** The other steered wheel is driven directly via a tie rod or track rod from the steering box drop arm. (2) Any passive arm or lever that serves primarily to retain the geometry of a mechanical system, rather than to transmit load or effort.

idling Of an engine, running at idle speed or ***tick over.***

idling jet See ***slow running jet.***

idling shaft (1) A ***layshaft*** or ***countershaft.*** (2) Shaft of an idler wheel or gear.

IFS Independent front suspension. See ***independent suspension.***

igniter Electronic ignition module that acts as a ***voltage booster*** or amplifier.

ignition Initiation of combustion. See also ***compression ignition engine***; ***ignition system***; ***spark ignition***.

ignition coil Induction coil or voltage transformer that provides the high tension voltage for the spark in spark ignition engines.

ignition delay Time interval between the spark and the initial release of heat energy of combustion in a spark ignition engine or between the start of injection and the start of ignition in a ***diesel engine***. Also ***ignition lag***.

ignition inhibitor Switch or switching circuit that prevents activation of ***starter motor***, for example when a vehicle with automatic transmission is in other than ***neutral*** or ***park***. Also ***neutral start switch***.

ignition lag See ***ignition delay***.

ignition switch (1) Electrical switch by which the ignition system of a vehicle is caused to function. Usually operated by a key which can also be turned to complete the ***starter motor*** circuit to initiate starting. (2) Also, and incorrectly, used to describe a key used to start and stop a diesel engine.

ignition system Electrical system devised to produce accurately timed sparks at the spark plugs of an engine, and consisting of a ***battery***, ***induction coil***, ***capacitor***, ***distributor*** or ***module***, ***spark plugs*** and the relevant switches and wiring. An alternative ignition system uses the ***magneto***. See Figure I.1.

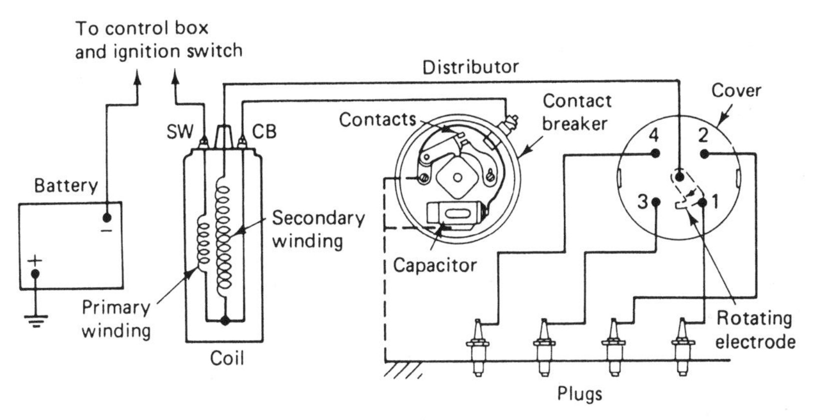

Figure I.1 Items in a conventional (non-electronic) ignition system.

ignition timing Timing of the spark relative to piston top dead center in a spark ignition engine. Usually expressed in degrees of advance. See ***advanced ignition***; ***retarded ignition***.

IHP See ***indicated horsepower***; ***indicated power***.

IMEP See ***indicated mean effective pressure***.

impact (1) Collision between a vehicle and other vehicle or other object, static or moving. (2) Any contact between solid objects where there is a change of relative velocity, and kinetic energy is lost, as with a hammer striking a nail.

impact horn Conventional type of electric horn for acoustic warning, in which a flexible diaphragm is struck or impacted by an electric armature.

impeller (1) The power input member of a torque converter. See Figure T.3. (2) The gas driven or driving rotor of an exhaust gas turbocharger. (3) Rotating member of centrifugal ***water pump***.

impingement cleaner ***Air cleaner*** or ***filter*** in which elements of woven wire are coated with a viscous covering, such as an oil from which trapped particles can be removed by washing.

in-car entertainment Sound reproduction system consisting of, for example, radio, tape or compact disc player specially devised to facilitate operation by the driver, and to be unaffected by vibration. Also television and video in goods vehicle ***sleeper cabs*** and passenger coaches. Also ***ICE***. (Informal)

in-line engine Engine with all cylinders in one plane of the crankshaft axis.

in-line fuel injection Fuel injection system in which each engine cylinder is provided with fuel by one dedicated pump/plunger of an in-line fuel-injection pump. See also ***distributor pump***.

in-line power steering Power-assisted steering system in which the powered steering effort is applied within the ***steering box*** or ***rack***, forming an integral powered unit. See also ***offset power steering***.

inboard brakes Brakes located close to the vehicle centerline rather than at the wheel hub, the retarding torque being transmitted to the wheels by way of the ***axle shafts*** or ***half shafts***. See also ***outboard brakes***.

inboard starter Inertia or Bendix type starter motor on which the drive pinion moves toward the motor body to engage.

inclination angle Of a wheel, the angle of the ***wheel plane*** to the vertical axis, according to the chosen axis system. See also ***camber angle***.

inclined engine In-line engine in which the cylinders are set at an angle to the vertical. Also ***sloper***. (Informal)

independent retarder A ***retarder***, the operation of which is separate from the main or ***service braking*** system.

independent suspension Suspension system in which the deflection of one wheel is not directly transmitted to the other, as in ***independent front suspension*** (IFS) and ***independent rear suspension*** (IRS). See Figure I.2.

indicated horsepower See ***indicated power***.

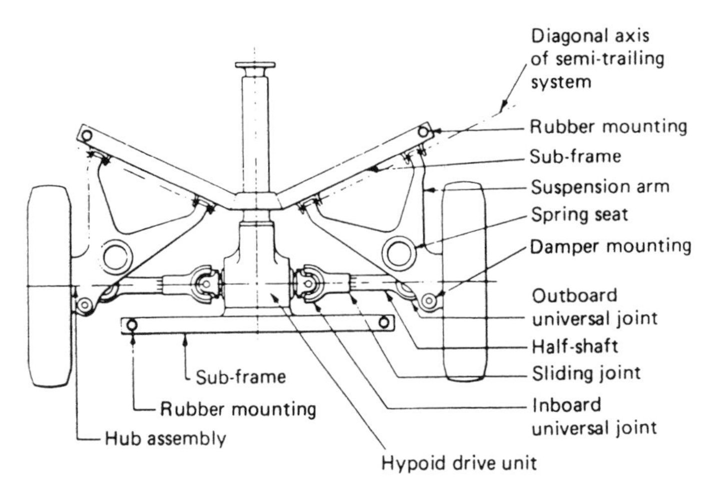

Figure I.2 A semi-trailing arm independent rear suspension.

indicated power (1) Engine power calculated from an indicator diagram. (2) Power developed in the cylinders rather than at the shaft. The sum of brake power and power lost to friction and pumping. See ***pumping losses***.

indicated thermal efficiency Ratio of indicated work available at the piston to the ideal work available from combustion. Effectively the thermal efficiency assuming zero friction and pumping losses. See also ***brake thermal efficiency***.

indicator (1) Instrument for visually recording engine cylinder pressure during a working cycle. (2) Direction indicator or turn signal lamp. See also ***trafficator***. (UK informal)

indicator diagram Mechanically or electronically produced map of cylinder pressure plotted against stroke or degrees of rotation for the power stroke or complete cycle of an engine. See also ***indicator***.

indirect injection Fuel injection into a ***prechamber*** or cell, in which ignition is initiated before the burning mixture enters the main combustion chamber. Often shortened to ID. See also ***air cell***; ***direct injection***; ***Comet head***; ***Lanova air cell***; ***prechamber***.

indirectly controlled wheel In an ***anti-lock system***, a wheel in which the braking action results from the signal provided by a sensor in another wheel. See also ***directly controlled wheel***.

induced drag Drag resulting from the generation of a vortex system as a result of the vehicle's tendency to lift, and therefore related to the body shape of the vehicle (or aerofoil) and its angle to the incident airflow. Increasing the angle will normally increase induced drag whereas profile drag will be substantially unchanged. Also ***vortex induced drag***.

induction Drawing in, as of air or an air/fuel charge into an engine. See also ***aspiration***.

induction air Air drawn into a working engine, before mixing with fuel.

induction manifold See ***intake manifold***.

induction port See ***intake port***.

induction stroke The stroke of the piston in an IC engine in which working fluid is drawn into the cylinder. Also ***intake stroke*** and ***suction stroke***.

induction system That part of a spark ignition engine in which the fuel and air are mixed and brought into the combustion chamber, including for example ***air filter/cleaner***, ***carburetor*** or ***fuel injection system***, ***intake manifold***, ***pressure charger***, ***intake port*** and ***valves***. In a diesel engine the fuel system would not normally be considered part of the induction system.

inductive kick Phenomenon which augments the secondary voltage in an ***ignition system*** by interrupting the current after ***points*** have been opened.

industrial tractor A vehicle other than an agricultural tractor, and for off-highway use or for use in highway construction and repair. Official definitions may limit weight and speed.

inertia braking Brake application on a ***trailer*** brought about by the towed vehicle moving up on the towing vehicle.

inertia disc Flywheel-like ***acoustic damper*** appended to rotating shafts to detune resonances and so reduce noise generation in specific frequency bands. See also ***tuned absorber***.

inertia drive See ***Bendix drive***.

inertia pinion See ***Bendix drive***.

inertia reel Seat belt with reeling mechanism that locks to provide constraint on sudden deceleration.

infrared gas analyser Instrument for quantitative measurement (and identification) of exhaust gases by absorption of infrared and related radiation. Also ***IRGA***. See also ***non-dispersive infrared***.

inhibitor Substance which prevents or slows down an undesirable reaction, such as the oxidation of a ***lubricant***, or corrosion of metal parts.

injection pump Device that supplies fuel under pressure to the injector of a fuel injection system. See also ***continuous spray pump***; ***distributor pump***; ***jerk pump***.

injector Device for introducing fuel under pressure into either the intake or combustion system of an engine. See also ***fuel injection***; ***injector nozzle***; ***pintaux nozzle***; ***pintle***. See Figure I.3.

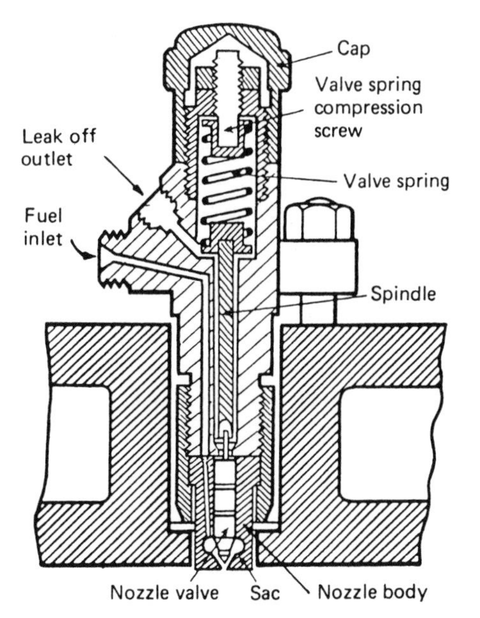

Figure I.3 A single nozzle diesel fuel injector.

injector nozzle Fine sprayer or atomizer through which fuel is injected into an engine. See also ***pintaux nozzle***; ***pintle***.

inlet valve See ***intake valve***.

inner dead center The equivalent of bottom dead center in a (horizontally or vertically) ***opposed cylinder engine***. Infrequently used as equivalent to BDC of an in-line or single cylinder engine.

inner liner Low diffusion (low permeability) covering of the inside of a tubeless ***tire*** to prevent air pressure loss.

inner tube Sealed rubber toroid which seats on a ***wheel rim*** and transmits the pressure of inflation to the ***tire carcass***.

inset dish wheel Wheel of which the ***nave*** face is outboard of the tire centerline. See Figure W.5.

installed power Power of engine with ancillaries, such as the ***generator***, being driven. The ***net power***.

instantaneous piston speed Piston speed at any specified point in its stroke or crankshaft angle. See ***average piston speed***; ***piston speed***.

instantaneous suspension center Projected geometric center of movement of suspension in bump or rebound mode at a stated point.

instrument panel Panel on which a vehicle's instruments are mounted. Also ***dash panel*** and ***dashboard***, particularly if instruments are mounted directly onto this item. See also ***binnacle***; ***fascia***.

insulated body (1) Thermally insulated body of vehicle intended for the conveyance of goods at low temperatures, such as foodstuffs. See also ***reefer***. (2) Tipper body with insulated sides and floor to prevent heat loss when conveying hot materials such as asphalt.

intake depression Mean reduction in static pressure below ambient in an engine air intake system, usually measured adjacent to the flange of the engine manifold or turbocharger.

intake manifold Manifold which distributes working fluid to intake ports. Also ***induction manifold***. See also ***induction system***; ***manifold***.

intake port Passage through which the induced air or air/fuel mixture passes to the inlet valve of an engine. Also ***induction port***; ***intake port***.

intake stroke See ***induction stroke***.

intake valve Valve that controls the admission of working fluid into the cylinder of an engine. Also ***inlet valve***.

integral body construction Form of construction in which there is no separate chassis, suspension and drive loads being reacted through the panels and structure of a torsionally stiff body. A ***monocoque*** structure. Also ***unitary construction***.

integrated retarder A ***retarder***, the operation of which is linked to that of the ***service brake***.

inter-axle differential Differential linking ***tandem axles***, or between the axles of a four-wheel-drive vehicle. See also ***transaxle***; ***transfer box***.

intercooler Heat exchanger that removes heat from pressure charged air. In strict (and original) usage, an intercooler was introduced between stages of compression. Also ***aftercooler***.

intercooling Cooling of pressurized air from a ***supercharger*** prior to admission to the cylinder. Cooling increases the mass of air induced.

interference angle Minimum apex angle between two inclined planes over which a vehicle can pass without striking any part of the vehicle underbody, usually measured as the included angle, for example 140 degrees. See also ***ramp angle***. See Figure L.3.

interleaf cushion spring Flat section spring that provides axial cushioning and therefore smoother engagement of the friction lining of a clutch. See also ***torsional drive spring***.

interlock In a change-speed gearbox, a device that prevents the engagement of two gears simultaneously. See also ***gate***.

intermediate gear Any gear between top and bottom gear.

intermediate plate See ***interplate***.

intermediate rod See ***relay rod***; ***track rod***.

intermediates Lightly treaded tires for racing in wet or dry conditions. (Informal)

internal combustion engine Engine in which energy is provided by combustion within a working chamber causing direct mechanical displacement of a piston, rotor, turbine, or other mechanical element. The ***gasoline (petrol) engine***, ***diesel engine***, ***Wankel engine*** and ***gas turbine*** are internal combustion engines. Often abbreviated to IC engine. See also ***external combustion engine***.

internal-expanding brake A drum brake with internal shoes.

interplate Disc that separates the two driven plates of a twin plate clutch. Also ***intermediate plate***.

intumescent Capable of swelling, as under the effect of heat or moisture.

IRS Independent rear suspension. See ***independent suspension***.

iso-octane Hydrocarbon used as a primary reference fuel in determining the octane rating of fuels, having assigned values of RON and MON of 100. See also ***Motor Octane Number***; ***Research Octane Number***.

ISO Viscosity Numerical viscosity system emanating from the International Organisation for Standardization, in which ***viscosity*** is indicated by viscosity bands in centistokes.

J

jack Portable or mobile device for lifting a vehicle or one side of a vehicle.

jackknife Loss of control in an ***articulated vehicle*** in which the tractive unit rotates about its vertical axis, even until it contacts the ***trailer***. The jackknife usually results from the locking and subsequent sliding of the tractor driven wheels, and should not be confused with ***trailer swing*** or ***trailer sway***. See Figure J.1.

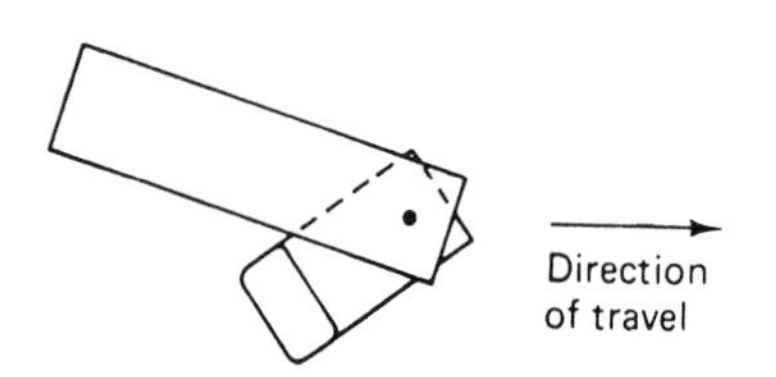

Figure J.1 Jackknife of an articulated vehicle.

jack rabbit start See ***stoplight drag***. (US informal)

jack-up Lifting of a vehicle body on cornering, particularly where instigated by a suspension geometry with a high roll center. Also ***jacking***.

jacking See ***jack-up***.

jacking bracket Reinforced mounting point under a vehicle by which the vehicle can be raised by a ***jack***, as for service or inspection. Also ***jacking point***.

jackshaft (1) Small shaft within a machine for transmitting rotary motion, as for example to a distributor. (2) A rotating shaft joining two other shafts. See also ***quill drive***.

Jake brake See ***engine brake***. Contraction of manufacturer's trade name.

jam switch See ***jamb switch***.

jamb switch Pushbutton light switch operated by opening and closing of door or hatch, and normally located in the jamb of the door. Also, but incorrectly, ***jam switch***.

Jaray car Car body shape derived from a combination of aerodynamic "teardrop" shapes, and characterized by a rounded front end and wedge-shaped tail form, after the Austrian aerodynamicist P. Jaray. See Figure J.2.

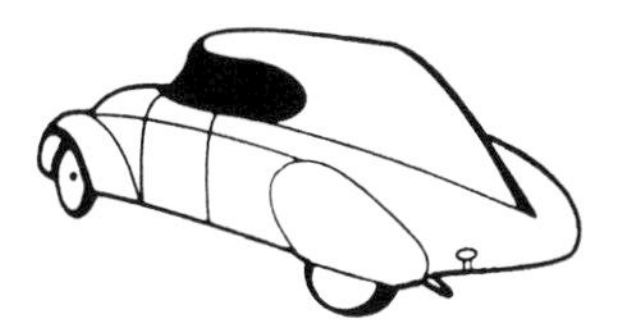

Figure J.2 Characteristic shape of a Jaray car.

Jeantaud steering Steering system in which the projected axes of the steered wheels meet at the projected axis of the ***back axle***, though at only one ***steer angle***. A geometric property also patented by Ackermann, by whose name the system is usually known in English-speaking countries.

Jeep dolly Undercarriage dolly with its own intermediate ***fifth wheel***, which prevents the overloading of the tractor unit under abnormal loading.

jerk pump Diesel fuel injection pump. Informal name for the cam-operated ***in-line plunger type*** pump.

jet An accurately drilled hole through which liquid can pass at a controlled rate, as in a ***carburetor***. See Figure C.3.

jounce Bump travel of a wheel suspension. (US)

journal Part of a rotating shaft that is supported by a bearing.

joystick control Hand control in form of a raised lever with two degrees of freedom, so that it can be moved backwards and forwards, and side-to-side, or a combination of both. So called because of the similarity of action to an aircraft's pitch and roll control lever of the same name.

judder (1) Low-frequency vibrations from brake or clutch assembly, the frequency being related to rotational velocity. (2) See ***clutch judder***. Also ***clutch shudder***.

juggernaut A large commercial vehicle. (Popular slang)

jump lead (US: jumper cable) Heavy-duty electrical leads whereby a vehicle with a discharged battery may be connected to an external charged battery, as for example the battery of another vehicle.

jump start To start the engine of a vehicle with a discharged battery by using jump leads.

jumper cable See ***jump lead***.

K

K-back See ***Kamm-back***.

Kadenacy effect In two-stroke motors, the creation of a partial vacuum by the sudden release of exhaust gases through exhaust ports. A means of improving exhaust ***scavenging***. After researcher Michael Kadenacy.

Kamm-back Body form termination of a vehicle in which a gentle taper is followed by an abrupt change to a flat back, designed to reduce vortex formation, after the German aerodynamicist W. Kamm. See Figure K.1. Also ***K-back***; ***K-shape***; ***Kamm tail***.

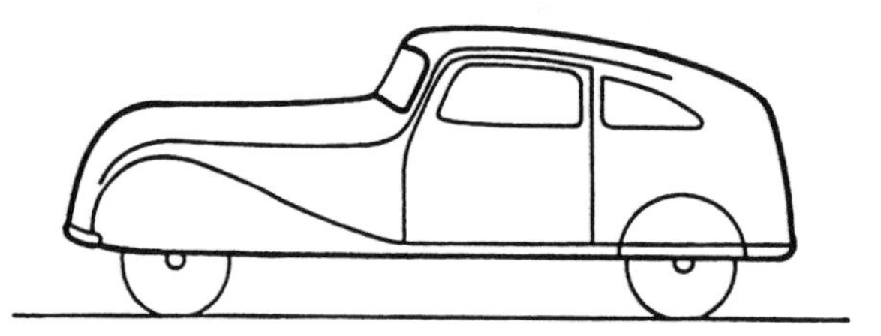

Figure K.1 A K-back or Kamm-back car of 1938.

kei Lightweight passenger car not exceeding 3.3 m length and 1.4 m width. From Japanese word meaning "light." Pronounced like K.

kerb weight (US: curb weight) Weight of an unladen commercial vehicle including fuel, water and oil, but subject to local definition.

kerbing rib See ***curb rib***.

kerf Continuous, and intentional, cut or slot in a tire ***tread***. See also ***sipe***.

kerosene Light petroleum distillate fuel consisting mainly of paraffins and isoparaffins, and of low ***octane rating***. Mainly used for gas turbines, but also as a low pollution diesel fuel. Also kerosine.

Kettering ignition system Commonly used inductive ***ignition system***, comprising ***induction coil***, ***breaker contacts***, ***capacitor*** and ***battery***, after originator Charles Kettering.

kick-up pipe See ***tail pipe***. (US informal) See Figure E.2.

kick start Pedal lever and ratchet mechanism for starting an engine, as for example a motorcycle.

kickdown (1) System that enables a driver to select a lower gear than the one automatically engaged by an automatic transmission, for example when accelerating rapidly. Operated by depressing accelerator pedal fully. (2) The act of using the kickdown facility. Also ***forced downshift***.

kingpin (1) Vertical or inclined shaft about which a steered wheel assembly pivots. Also fulcrum pin; knuckle pin; pivot pin. (2) Main fastening member between a ***fifth-wheel*** and ***semi-trailer***. See Figure S.8.

kingpin angle See ***kingpin inclination***.

kingpin axis See ***steering axis***.

kingpin centers Transverse distance between intersection points of ***kingpin axes*** and ***steered wheel axes***.

kingpin inclination (1) Angle of inclination of ***kingpin axis*** to vertical longitudinal plane. (UK) (2) Angle in front elevation between the steering axis and the vertical. Also ***kingpin angle***; ***swivel angle***; ***steering axis inclination***. Terminology remains although true kingpins are now rarely used in passenger cars. See Figure C.2.

kingpin offset (1) At ground, the horizontal distance in front elevation between the point where the ***steering axis*** intersects the ***ground plane*** and the center of tire contact. (2) At wheel center, horizontal distance in front elevation between wheel center and ***steering axis***. See Figure C.2.

knee bolster Impact absorbing and protective panel below ***dash panel*** which absorbs impact with the knee in the event of frontal impact while minimising risk of injury.

knock (1) Noise resulting from the spontaneous ignition of a portion of the air-fuel mixture in an engine cylinder and occurring ahead of the normal spark-initiated advancing flame front. (2) Detonation of the fuel mixture in an engine cylinder or the noise thereof. See also ***detonation***; ***diesel knock***; ***ping***; ***pinking***; ***pre-ignition***; ***runaway knock***; ***spark knock***; ***surface ignition***.

knock-on wheel Quickly detachable wheel typically with a splined center fitting a splined hub and secured by a single nut, tightened by a hammer. See also ***Rudge nut***.

knock-out axle Detachable axle usually of a ***low-loader***, removal of which allows further lowering of the loading bed.

knock rating Octane rating. (Informal)

knock sensor Instrument which detects the onset of detonation in an IC engine. See also ***Motor Octane Number***; ***Research Octane Number***.

knuckle pin See *kingpin*.

Krypton test Proprietary instrumented diagnostic and performance test for IC engines.

Kumm transmission Constantly variable ratio transmission using a flat belt and expanding pulleys.

L

L-head engine (UK: side valve engine) In-line engine in which inlet and exhaust valves are on the same side of the cylinder block and set within the block, with poppet valves stem downwards. (Obsolete for most automotive uses) See also ***F-head***; ***I-head***; ***side valve***. See Figure S.5.

labizator A (usually torsional) steering and front suspension compensating spring of Z form in the road plane. Also ***Z-bar***.

lacquer Smooth deposit on engine parts caused by polymerisation of decomposition products of ***fuel*** and ***lubricant***.

ladder chassis Chassis in which parallel side members are joined at intervals by transverse beams, giving the appearance of a ladder.

lag Angle after bottom dead center at which an engine (inlet) valve closes. See also ***lead***.

lambda sensor Electrochemical sensor that relays data on oxygen content of exhaust gases to an electronic engine management system or to laboratory equipment, and thus enables corrections to be made for divergence from ***stoichiometric*** mixture relationship.

laminated glass Safety glass in which a transparent plastic film is sandwiched between plates of glass, thus reducing splintering and resisting penetration by occupant. See also ***toughened glass***.

lamp Lighting unit consisting of lens, reflector, filament or light source and housing.

land implement Towed implement for use in agriculture, forestry, dredging and other specialist duties. (Mainly official usage)

land locomotive (1) Effectively, a land tractor exceeding official land tractor weight. (2) An agricultural (steam) traction engine.

land tractor Tractor mainly for use in agriculture, forestry and related operations. Official regulations may limit weight and on-highway usage.

landing gear Retractable or removable supports for the front end of a ***semi-trailer***, sometimes with small wheels to provide limited mobility. Also ***landing legs***. See also ***dolly***; ***sandshoe***. See Figure L.1.

landing legs See ***landing gear***.

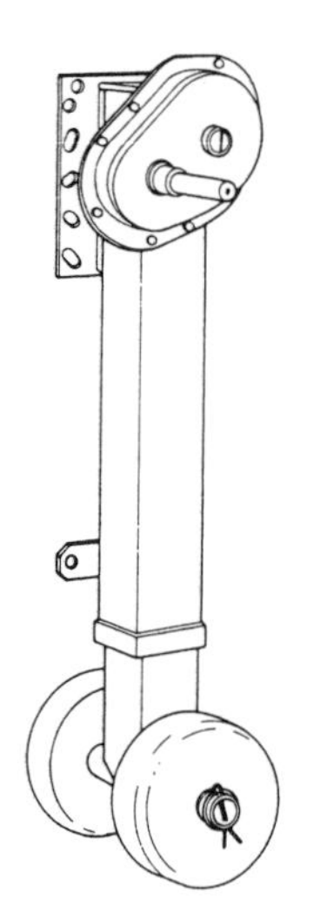

Figure L.1 A leg of a semi-trailer landing gear.

Lanova air cell Narrow throated cavity located in the head of certain types of diesel engine, normal to the cylinder axis and on the opposite side of the combustion chamber to the injector where it entraps and stimulates ignition of part of the injected charge. The Lanova air cell is not strictly a pre-combustion chamber.

Lanova head Cylinder head incorporating a Lanova air cell.

lap and diagonal belt Active occupant restraint system in which a continuous fabric belt is fastened normally to near the centerline of a vehicle to provide horizontal restraint at hip level and diagonal restraint from outer shoulder across the thorax.

lap belt Safety belt that affords constraint on frontal impact across the occupant's lower waist or hip.

lash See ***play***.

lashing eye Attachment lug with eye or flanged hole for securing by rope a vehicle in transit to a transporter or other shipment vehicle.

lateral acceleration The "acceleration" that gives rise to the sideways force on a vehicle (and its occupant) when cornering.

lateral control force See ***cornering force***.

lateral runout Oscillation of the plane of symmetry of a rotating road wheel as a result of static misalignment. See also ***shimmy***, which is a dynamic oscillatory condition, and ***waddle***. ***Wheel-wobble***. (Informal)

Layrub joint Proprietary universal joint using rubber bushes mounted on an intermediate carrier. Also ***Layrub coupling***.

layshaft (US: countershaft) Shaft in a gearbox running parallel to the main shaft and carrying the paired gear wheels or pinions that effect the changes in gear ratio.

LDC See ***bottom dead center***.

lead (1) Tendency of a vehicle to deviate to left or right, for example as a result of steering misalignment, asymmetric loading or inequality of tire pressures. (Pronounced as leed) (2) Lead compound ***anti-knock additives*** for gasoline. (Informal) (Pronounced as led) (3) Angle in advance of ***top dead center*** at which an engine induction valve opens. (Pronounced as leed) See also ***lag***. (4) Metal element of high specific gravity, and high malleability, gray in color, used in ***lead-acid batteries***.

lead-acid battery Battery consisting of lead-acid cells in series, normally 12 volts for passenger cars and light commercial vehicles.

lead alkyl Lead compounds such as ***tetraethyl lead*** and tetramethyl lead which act as ***octane improvers*** in gasoline.

lead-free Of gasoline, containing no lead-compound ***anti-knock*** additives, or lead in any other form.

lead-on ramps Guides located behind the ***fifth wheel*** of an articulated tractor to guide the ***semi-trailer kingpin*** into engagement for towing.

leading arm Suspension linkage which supports the wheel axle on a forward facing sprung lever.

leading shoe (US: primary shoe; forward shoe) Shoe of a ***drum brake*** system in which the actuated end leads, facing the normal direction of rotation. See also ***twin leading shoe***. See Figures C.1 and D.8.

leaf spring (1) Spring built from superimposed narrow, flat sectioned plates or blades which resist load in bending. See Figure L.2. (2) Spring consisting of one tapered member in bending, also called ***single leaf***. See also ***quarter elliptic***; ***semi-elliptical leaf spring***. See Figure A.6.

leaf spring strap Strap, usually of steel and lined with a resilient sleeve, for holding together the leaves of a leaf spring, while permitting relative motion of the leaves. See also ***clinch***, which is an alternative device. See Figure L.2.

lean-burn engine Engine capable of running on a fuel-air ratio that is significantly lower than ***stoichiometric***.

lean mixture Inducted air/fuel mixture containing an excess of air. A ***weak mixture***.

leaned Having the ratio of air to fuel in an inducted mixture increased, so that the charge is weaker. (US informal)

left-hand drive Driving position on left of vehicle, as generally adopted in countries where vehicles drive on the right.

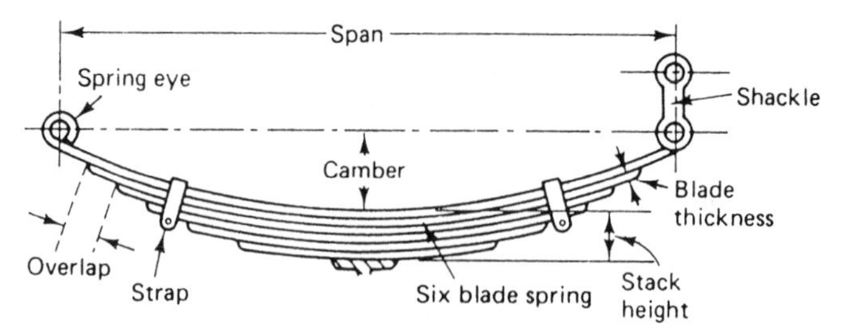

Figure L.2 Semi-elliptical leaf spring nomenclature.

Lemoine Stub axle and kingpin configuration for ***beam axles***. See also ***Elliot axle***.

lever type shock absorber Suspension vibration damper in which the suspension input is reacted by lever arm from a chassis-mounted hydraulic damping unit.

license plate lamp (UK: registration plate lamp) Lamp to illuminate the license plate at the rear of a vehicle.

lift (1) Of a ***poppet valve***, the distance by which the valve is raised from its seated position when fully opened. See ***cam***. (2) Of a wheel, the difference between the high and low point of the radius. The ***radial truth***. (Obsolescent term) See also ***wobble***.

lift axle See ***lifting axle***.

lift gate (UK: tail lift) Power-operated tailgate, particularly of van or box-body commercial vehicle, which converts into a platform whereby cargo can be lifted from street level to vehicle cargo floor level. Also ***lift tail gate***.

lift off tuck-in Tendency of some suspension configurations to abruptly apply toe-out on curve when foot is lifted off the accelerator, thus effectively tightening the radius of turn, if rear-wheel drive.

lift-over height Height from ground plane of the ***sill*** or other permanent obstruction over which an object, such as baggage or cargo, has to be lifted for stowing on a vehicle.

lift pump Low-pressure feed pump transferring fuel from tank to ***carburetor*** or ***fuel injection pump***.

lift tail gate See ***lift gate***.

lifter (1) Actuating mechanism that lifts or opens a ***poppet valve***. (2) A valve lifting tool. (3) A ***tappet***. (US informal) See also ***hydraulic lifter***; ***roller lifter***.

lifting axle Axle of a ***tandem axle*** undercarriage with mechanism for raising above ground contact when the vehicle is unladen or lightly laden. Term also applies to the complete bogie suspension of which an axle can be lifted. Also ***lift axle***.

light (1) Window in a vehicle passenger compartment. Also used in context of the number of windows, for example six-light, being a passenger car with three separate windows on each side. See also ***backlight***; ***quarter light***. (2) Beam of light given off by a lamp (but not an alternative word for lamp).

light locomotive A road locomotive with an unladen weight nominally between 7370 kg and 11, 690 kg. (Mainly UK usage)

light-off Coming into operation of a ***catalytic converter*** when operating temperature is reached.

light-off time Time from engine starting to operating temperature of an exhaust ***catalytic converter***. Also ***LOT***.

light van/truck Vehicle of restricted weight (typically under 3 tons in the UK, or as specified by the regulating authority), which may be legally driven by a person not holding a commercial vehicle licence.

limit cycle control Closed-loop engine control system in which a feedback signal is sent only when a pre-set limit of the measured parameter is exceeded. See also ***proportional control***.

limited slip differential Differential in which the difference in rotational speed or torque between the two output shafts is mechanically limited to prevent wheel spin on difficult terrain. Mainly used on vehicles where an off-highway capability is important, or on commercial vehicles, competition cars and passenger cars to improve ***traction*** on ice and snow. See also ***locking differential***.

liner (US: sleeve) Hard metal insert in engine ***cylinder block***, in which piston runs. (UK) Also ***cylinder liner***. See also ***dry liner***; ***sleeve valve***; ***wet liner***.

lining (1) Friction material attached to the shoe of a drum brake. (2) The pliant and decorative surface of a vehicle interior, as in ***headlining***.

linkage power steering Power-assisted steering system in which a conventional manual system is aided by hydraulic (or pneumatic) effort applied directly to a steering linkage such as an ***idler arm*** or ***track rod***. See also ***offset power steering***.

liquefied natural gas Natural gas as an automotive ***alternative fuel***, principally ***methane***. Also ***LNG***. See also ***compressed natural gas***.

liquefied petroleum gas An automotive ***alternative fuel*** consisting mainly of propane, sometimes with the addition of butane, which can be stored at relatively low pressure. Also ***LPG***. Also ***Autogas***, a trade name though extensively used as a generic term for LPG in automotive applications. See also ***liquefied natural gas***.

liquid-cooled Cooled by conduction and convection to the passage of a liquid, such as water. See ***cooling system***; ***radiator***.

lithium grease Grease based on lithium soaps, common in automotive applications.

live axle Axle that transmits power to a pair of wheels. See also ***fully-floating axle***; ***non-floating axle***; ***semi-floating axle***.

LNG See ***liquefied natural gas***.

load distribution (1) Load apportioned to each axle of a vehicle. (2) The ratio of load distribution between axles.

load floor The cargo-carrying floor of a commercial vehicle or trailer. See also ***cargo floor***; ***bottom board***.

load overhang See ***overhang***.

load proportional brake control System or device that regulates the input force to the brakes on an axle in proportion to the load on that axle.

load sensing valve Valve that adjusts the brake performance of a vehicle to the axle load, as for example to reduce risk of locking of rear wheels of an unladen ***semi-trailer***.

load space lamp Door-actuated illumination of baggage compartment of a car. Also ***boot lamp***. (UK)

load transfer Effective increase or decrease in axle load due to acceleration or braking, or lateral transfer across axle on cornering.

loadbase Distance between the centerline of the front wheels of a commercial vehicle and the transverse centerline of the load if evenly distributed, or the center of gravity of its load. See Figure L.3.

loader A predominantly off-highway, earthmoving vehicle equipped with a loading shovel for gathering up and lifting spoil and loose material for loading onto, for example, a ***tipper truck***. See ***loading shovel***. See also ***backhoe loader***.

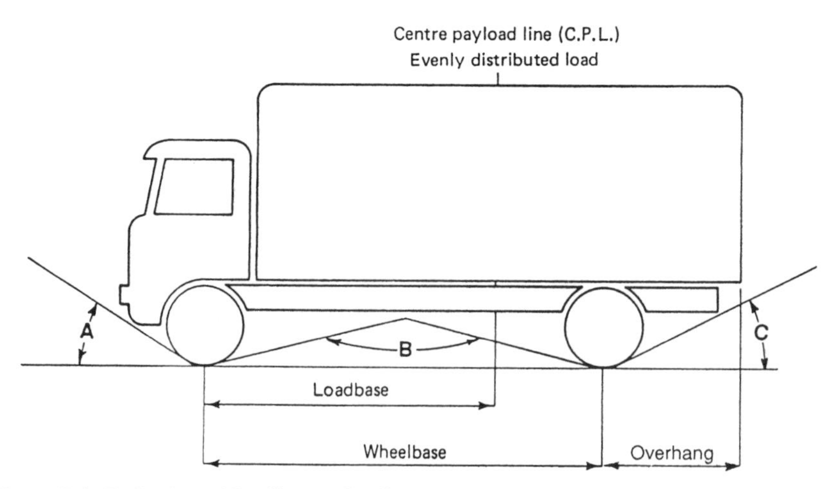

Figure L.3 Definition of loadbase, wheelbase, overhang, approach angle (A), interference angle (B), and departure angle (C).

loading shovel Usually hydraulically operated bucket or shovel for gathering material from a site, lifting it, and loading it, as for example onto a hopper or tipper truck for removal.

lobe Part of a profile cam that extends beyond the base circle, and that brings about the lift of a follower or tappet.

lock (1) Angle of rotation about steering axis of steered wheels. See also ***angle of lock***; ***steering angle***. (2) Maximum angle to which steered wheels can be turned, as in context of lock-to-lock. (3) Mechanism for securely closing a door or hatch, or for otherwise ensuring security, as against theft.

lock actuator Device to keep ***parking brake*** applied independently of air pressure in an air brake system.

lock synchronizer Change-speed gear ***synchronizer*** mechanism that positively locks paired gears, usually by means of a dog clutch. The conventional synchromesh mechanism. See Figure S.13.

lock-up Direct drive mode in which the driving and driven elements of a torque converter are mechanically coupled so that the complete unit, with casing, rotates at engine speed, thus eliminating fluid friction losses in the converter and increasing fuel economy. Also ***lockup***.

lock-up clutch Clutch that directly couples the ***torque converter*** output to the engine drive in an ***automatic transmission***, as in direct drive mode. See Figures A.5 and L.4.

locked wheel Non-rotating wheel of a moving vehicle traversing the road surface without rotation, as in a ***skid***.

locking differential Differential with facility for mechanically locking together both ***half shafts*** of an axle, the differential action no longer functioning. Particularly used on heavy vehicles and off-highway vehicles for operation on rough or low-friction terrain. Also ***diff-lock***. (Informal) See also ***differential lock***; ***limited slip differential***. See Figure L.4.

locomotive Road tractive unit, normally for drawing multiple trailers. See also ***light locomotive***; ***heavy locomotive***; ***land locomotive***; ***road train***. (Terminology rarely used except in legal contexts.)

log bunk Transverse support for logs on a logging vehicle.

logging vehicle Vehicle for conveying logs and lumber. A ***logger***. (Informal) See also ***pole trailer***.

longitudinal slip (1) Tendency of a wheel's tire to slip slightly in normal drive or braking mode. (2) Ratio of ***longitudinal slip velocity*** to ***spin velocity*** of the straight, free-rolling wheel.

look-up tables See ***map***.

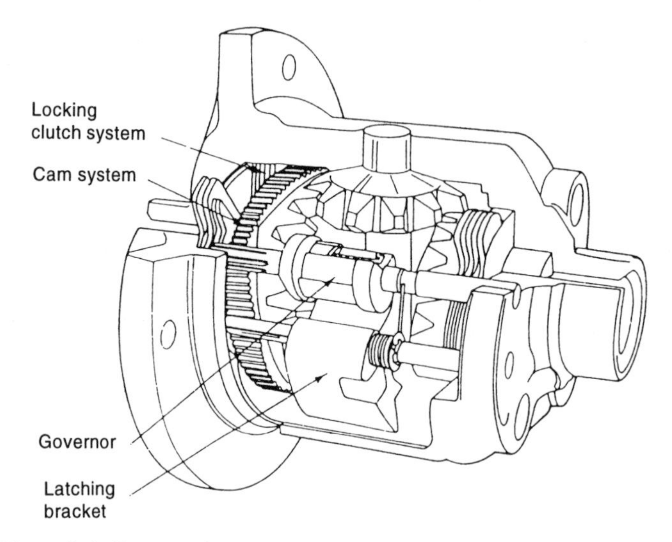

Figure L.4 Cutaway drawing of an Eaton locking differential.

loom Assembled wiring system of a vehicle, particularly as a unit prior to installation. Also ***wiring harness***.

loop-scavenging Scavenging of a ***two-stroke engine*** by vertical circulation within the cylinder and to the exhaust port of burned gases ahead of the incoming charge. See ***cross scavenging***; ***Schnuerrle system***.

lorry (US: flatbed truck) Heavy goods vehicle, particularly an open or platform truck. (Mainly UK usage)

louver Opening for ventilation consisting of parallel slats. Also ***louvre***, particularly in UK.

low-bed trailer (UK: low-loader) Open truck trailer constructed to provide a low platform height.

low-boy A ***low-bed trailer***. (US informal) Also ***lo-boy***.

low heat rejection engine See ***adiabatic engine***.

low loader Platform truck so equipped that the tilting of the platform or the lowering of an extension ramp facilitates direct loading, particularly of heavy mobile equipment, from street level. See also ***beaver tail***.

low profile Of low height in relation to width, but notably of tires with wide tread and shallow sidewall.

low sulfur diesel (UK: low sulphur diesel) Diesel fuel of reduced sulfur content, used particularly in environmentally sensitive areas.

lower beam (UK: dipped beam) Headlamp beam to illuminate the road ahead of a vehicle without causing undue glare to other drivers. Also ***meeting beam***. See also ***main beam***; ***upper beam***.

lower dead center See ***bottom dead center***.

LPG See ***liquefied petroleum gas***.

lubricant Substance used to reduce friction, such as an ***oil***, ***grease***, or solid such as ***graphite***.

lubricity Ability of a fuel, oil or grease to lubricate, particularly under heavy duty or ***boundary conditions***.

Lucar connector Proprietary, but widely used, electrical connector of the spade type.

lumber body Platform truck or trailer for the carriage of sawn lumber.

lug Un-siped block feature of ***tread*** pattern, usually of off-road tires.

lug-type terminal post Flat rectangular section terminal post to which mating lug is bolted.

luggage boot (US: trunk) See ***boot***.

lugging Ability, particularly of a working vehicle such as an off-highway vehicle, to pull through a temporary overload.

lugnut A ***wheelnut***. (US)

lurching Side-to-side oscillation of a vehicle without ***roll***.

luton Van in which the van body is of greater height than the driver's cab and extends forward above the ***cab***. See also ***box-van***; ***panel body***. See Figure L.5.

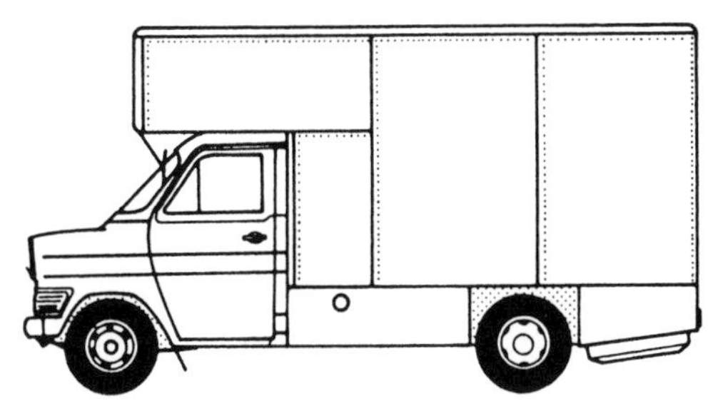

Figure L.5 A panel body luton van.

M

MacPherson strut Telescopic independently sprung suspension member incorporating ***damper***, and fixed at its upper end to the body shell or chassis, the lower end being located by linkages which counteract transverse and fore and aft movement. The original system used an ***anti-roll bar*** for longitudinal location. Usually incorporated in a steered front suspension system. See Figures A.3 and M.1.

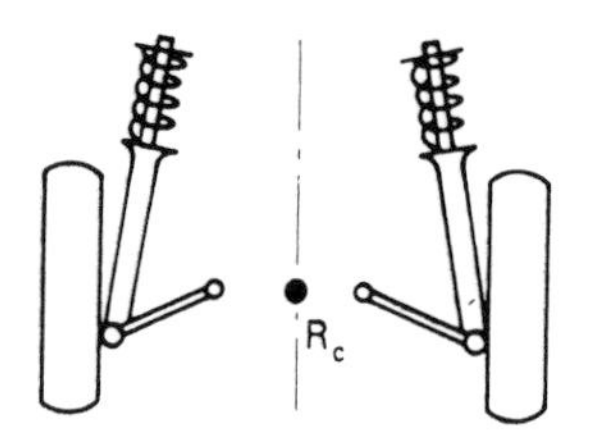

Figure M.1 MacPherson strut suspension, showing position of roll center.

MAF Mass air flow, particularly in relation to engine induction.

MAF sensor Telemetric device for measuring or monitoring the mass flow of air into an engine.

magnesium The lightest commercial structural metal with a specific gravity two-thirds that of ***aluminum*** but with generally lower resistance to atmospheric corrosion and acid attack. Easily machined (with precautions), magnesium is used mainly in cast alloy form.

magnetic plug Plug fitted in the ***oil pan*** or ***sump*** to collect ferrous debris.

main bearing Journal bearing that locates and supports the ***crankshaft*** in the ***crankcase***.

main beam (US: driving beam; high beam) Full beam of a headlamp, for use when no traffic is approaching. Also ***upper beam***.

main jet Principal jet in a ***carburetor*** through which the greater proportion of fuel flows in normal steady-state operation.

main metering system Of a ***carburetor***, the float chamber or other fuel source, the main ***discharge tube*** or ***jet*** and ***venturi***, excluding ***compensating***, ***idling*** and other systems.

manifold System of ducts or pipes that divides a flow and conducts it to more than one point of delivery or that unites a flow from a number of sources for delivery at one point. See ***exhaust manifold***; ***intake manifold***. See Figure P.5.

manifold absolute pressure The mean gas absolute static pressure in an engine induction manifold, usually measured by a mercury manometer. Also ***MAP***.

manifold depression Difference between the mean static pressure within an engine induction (intake) manifold and ambient pressure. Although the difference is usually negative (a depression) in a naturally aspirated engine, it will normally be positive in a supercharged engine. Use of the term manifold pressure avoids this anomaly. Used as a means of controlling ignition timing, and of augmenting brake pedal effort. See also ***servo***.

manifold pressure See ***manifold depression***.

manual brake system Mechanical or hydraulic brake system in which driver pedal effort is unassisted by ***servo***.

manual shift See ***manual transmission***.

manual steering Steering in which the effort of the driver is unassisted by mechanical power. See also ***power steering***.

manual transmission Transmission in which gears are changed by a driver-operated lever mechanism. See also ***column change***; ***gear lever (UK)***; ***stick shift***.

map Electronic data-bank from which signals controlling one or more parameters can be output in consequence of one or more signals (for example, engine revolutions and manifold vacuum) being input. Also ***look-up tables***. (Informal)

marker lamp Any lamp that indicates a vehicle's position, but particularly a light on a trailer or combination where front or rear lamps might not be seen. Also ***position lamp***. See also ***side marker lamp***.

Marles steering gear Proprietary ***hourglass worm and roller*** steering gear.

Marles-Weller steering gear Proprietary ***worm and peg*** steering gear.

masked valve Poppet valve recessed into its seat to give more efficient operation.

mast jacket Tube in which a steering column turns. (US obsolescent)

master cylinder (1) Primary source of pressure in a hydraulic system such as a brake or clutch system, containing the piston by which pressure is applied and connected to a source of hydraulic fluid. (2) Primary unit for dispensing hydraulic fluid under pressure in a hydraulic system. See Figure M.2.

Maxaret brake Proprietary anti-lock brake system, originally developed for aircraft but subsequently modified for road vehicles. (Obsolescent)

maximum brake power Maximum measured power of an IC engine, as measured on a ***dynamometer*** or brake.

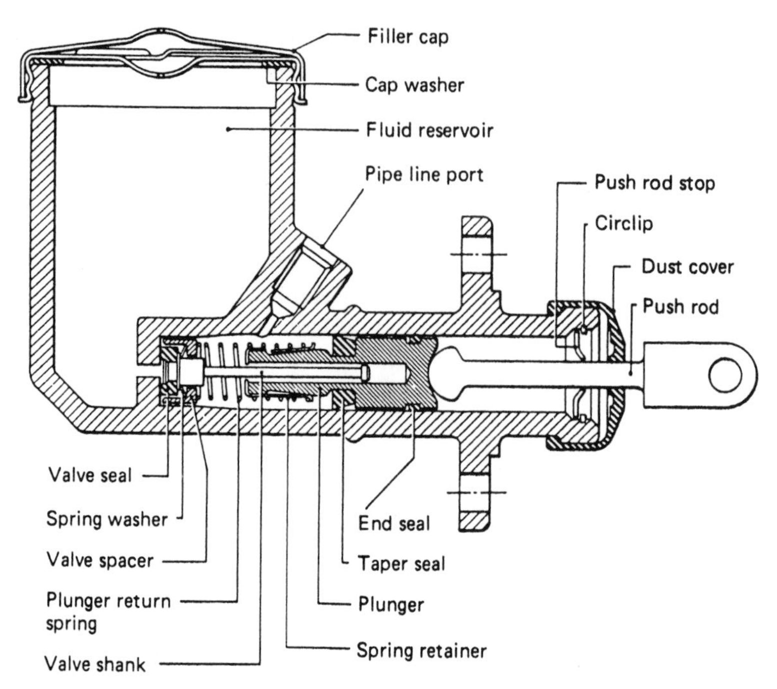

Figure M.2 A brake master cylinder.

mean effective pressure The mean pressure in an engine cylinder during the working or power stroke. See also ***brake mean effective pressure***; ***indicated mean effective pressure***.

mean piston speed Effective distance travelled by a piston per unit time, given by the formula S = 2LN, where S is the piston speed, L is stroke, and N is the number of revolutions per unit time. Also called ***mean piston speed***. Use of the unqualified term ***piston speed*** risks confusion with ***instantaneous piston speed***. Also ***average piston speed***.

mechanical trail See ***caster offset***.

mechanical transmission Transmission using mechanical rather than hydraulic or electrical principles.

medium-speed diesel Diesel engine with an operating speed in the range of 250 to 1000 rpm. Term mainly used in marine or industrial engine connotation. Most automotive engines are classed as high-speed.

meeting beam See ***dipped beam***; ***driving beam***. Also ***lower beam***.

Merit Rating System of rating engine deposits of combustion and lubrication, typically on a scale of 0 to 10, with 10 indicating complete cleanliness. See also ***Demerit Rating***.

metering rod Valve consisting of variable section rod and orifice by which flow of a fluid can be metered, as from the ***float bowl*** or chamber of a carburetor. Similar to a ***needle valve***.

methane Hydrocarbon constituent of natural gas, also produced by sludge digestion. An alternative fuel for spark ignition engines, though requiring storage in liquid state at low temperature.

methanol Methyl-alcohol as an ***alternative fuel*** or as a component of gasoline. It is not classed as an ***additive***.

methyl tertiary butyl ether Oxygen bearing anti-knock fuel blend component, no longer generally classed as an ***additive***. Also ***MTBE***.

Mexican overdrive Practice of coasting downhill with gears disengaged, particularly of a commercial vehicle. (Mainly US usage) Also ***Scotsman's sixth***.

micronic Of very small size, as would be measured in microns, as for example a micronic filter capable of trapping very fine particles. (Informal)

mid-engine (1) Engine located within or immediately behind the passenger compartment of a passenger car. (2) Engine located within the wheelbase and beneath the floor of a public service or commercial vehicle. See also ***forward control***; ***normal control***; ***rear engine***.

mild steel General-purpose grade of malleable and ductile low carbon steel, widely used for automobile bodywork and chassis construction, weldable and easily machined. Specific designations are currently preferred to the general term mild steel.

mileometer (US: odometer) Mechanical or electrical meter for indicating distance travelled.

Milner's Theory Theory for predicting life expectancy of a component operating at varying stress levels. See also ***Corten-Dolan Theory***.

mineral oil Light hydrocarbon distillate lubricating oil. Term usually distinguishes from vegetable oil, such as castor oil.

minibus Small bus, usually based on a van chassis, and usually equipped to carry less than 20 passengers.

misfire Failure of a fuel charge to fire or ignite in the proper way.

mixing chamber See ***barrel***. See also ***carburetor***.

moan Sustained low-frequency sound, particularly as generated by a ***tire***.

mobile crane Crane with on- and off-highway capability, and therefore equipped with steering, braking and lighting to meet appropriate legislative requirements.

modulated displacement engine Engine in which the swept volume can be changed to suit operational or environmental requirements.

module Semi-conductor control for electronic ignition circuit. (Informal) See ***electronic control module***. See also ***breakerless ignition***.

molybdenum disulphide (US: molybdenum disulfide) Black powdery solid sometimes added to ***lubricants*** to reduce friction.

MON See ***Motor Octane Number***.

monobloc construction Of a reciprocating engine, having the cylinders cast in a single block through which cooling water can flow continuously. Also used of an engine in which the cylinder block and crankcase are made as one unit.

monocar A single-seater car. (Archaic)

monochromatic lamp Lamp, the surface of which may be colored when unlit, as with body color, but which allows a white or colored light to show through when lit.

monocoque A structure, as for example a car body, consisting of a torsionally rigid shell. See also ***integral construction***. See Figure B.4.

monolith Originally a large block of stone, but in automotive parlance the catalyst-coated ceramic element of an exhaust ***catalytic converter***.

monomer Molecule of usually low molecular weight that, when bonded, comprises part of a ***polymer***.

moped A light motorcycle equipped with pedals for starting and assistance in hill-climbing.

Morse Test Engine test for multi-cylinder engines in which each cylinder of the running engine is stopped in turn so that its contribution to the total output can be measured. See also ***Motoring Test***.

MoT test Compulsory British annual vehicle condition survey, after the Ministry of Transport, the now superseded government authority that initiated the test.

motion shaft The mainshaft or layshaft (countershaft) of a ***gear train***, often expressed as first motion shaft, second motion shaft, etc. Probably originated in horology. (Obsolescent)

motion transmissibility Ratio of ***wheel hub velocity*** to ***tire footprint velocity***.

motor (1) A prime mover. The engine of a vehicle. (2) A motor car or automobile. (Mainly UK informal) (3) To drive or travel by car.

motor caravan (US: motor home) A rigid motorized vehicle, usually built on a van or light bus chassis, with residential leisure facilities. See also ***camper***.

motor home See ***motor caravan***. Also ***motorhome***.

motor method A standard test for fuel ***octane*** rating. See also ***research method***.

Motor Octane Number A guide to anti-knock performance of a fuel under relatively severe driving conditions, as at full throttle when inlet temperature and engine speed are both relatively high. It is derived from one of the standard comparative tests using the Cooperative Fuel Research (CFR) engine. Also ***MON***. See also ***Research Octane Number***.

motor scooter Motor bicycle with small wheels, and usually with leg shields and open foot platform formed as a unit to give weather protection and to facilitate mounting.

motor spirit Petroleum fuel for spark ignition motor vehicle engines. Also ***gasoline***; ***petrol***. (Obsolete except in legal usage)

motor stall torque Maximum torque that a motor can maintain for two cycles under specified conditions. (SAE definition)

motor tractor A motor vehicle for haulage rather than load carrying, nominally with an unladen weight below 7370 kg. (Mainly UK usage)

motorcycle Two-wheeled powered vehicle steered by handlebars and not equipped with pedals or other means whereby rider can assist motion. See also ***motorcycle combination***; ***moped***.

motorcycle combination Motorcycle equipped with a ***sidecar*** or ***buddy seat***, thus making the combination a three-wheeled vehicle.

Motoring Test Method of experimentally measuring mechanical losses in an engine. See also ***Morse Test***.

Moulton suspension Combined gas or rubber and fluid coupled suspension of type devised by A. Moulton. See also ***Hydrolastic suspension***; ***Hydragas suspension***.

mounting ring Fitting, with facilities for aiming adjustment, on which a headlamp or its light unit is mounted.

moving barrier A test barrier, sometimes mounted on a wheeled chassis, which can be moved at prescribed velocity to simulate impact with test vehicle.

moving van (UK: removal van; pantechnicon) Large-capacity van for removal of household effects. See also ***pantechnicon***.

MTBE See ***methyl tertiary butyl ether***.

mu-split Sudden change in road or ground surface friction, relevant to its effect on braking or cornering. Also ***μ-split***, using Greek character for coefficient of friction.

mudguard (US fender) Narrow curved panel mounted over or behind a wheel, to deflect tire-generated spray and roadstones.

mudwing Mudguard, particularly for the rear wheels of a commercial vehicle.

muff coupling Coupling consisting of two half sleeves keyed to a shaft and bolted together.

muffler (UK: silencer) Primary silencing element in an ***exhaust system***.

multi-fuel engine IC engine capable of running on more than one type of fuel, as for example LPG and gasoline.

multigrade oil Engine or gear oil which meets one or more relevant SAE viscosity grade classifications.

multi-leaf spring See ***semi-elliptical spring***.

multi-plate clutch Clutch with more than one driven plate. See also ***interplate***.

multi-point injection ***Fuel injection system*** in which the fuel is injected into each individual cylinder, rather than into the ***inlet tract***. See also ***downstream injection***; ***throttle-body injection***.

multi-pull brake Hand braking system, often fitted with ratchet mechanism, that requires more than one pull for full operation.

multi-spark ignition Ignition initiated by a rapid sequence of sparks between the ***spark plug electrodes***, rather than by a single spark.

multi-stage Of a gas turbine or turbocharger, having more than one stage of working blades.

multi-throw crankshaft Shaft with cranks set at equal angles of less than 180 degrees.

multi-way connector One-piece electrical connector for connection to two or more circuits. See also ***sub-harness***.

multiple-disc clutch See ***multi-plate clutch***.

mushroom valve Valve with narrow stem surmounted by a disc-shaped head, resembling an elongated mushroom. The conventional intake and exhaust valve configuration. ***Poppet valve*** is the favored term.

N

NACA duct Aerodynamically profiled shallow duct, developed by former National Advisory Committee for Aeronautics. (US) See also ***air scoop***.

nacelle A protective or streamlined cowling.

napthenic oils Lubricating oils with a high proportion of napthenic to ***paraffinic*** stocks. They have generally lower ***pour points*** and produce softer carbon residues if partially oxidized than do paraffinic oils.

narrow cut Petroleum fraction with relatively small difference between initial and final boiling points.

narrow V-engine Engine of V configuration with cylinder axes at less than 60 degrees to each other, and in some cases at an angle so small that the cylinders are staggered to avoid interference, as for example in some engines by Lancia.

naturally aspirated See ***normally aspirated***.

nave (1) Hollow or dished center part of a road wheel on which the rim is mounted. (2) The face of a wheel that is bolted to the ***hub***. See also ***nave face***. See Figure W.5.

nave face The flat surface of the nave that is bolted to the ***hub***.

NDIR See ***non-dispersive infrared***.

nearside Side of a vehicle normally nearest to the ***kerb (curb)***.

needle bearing Rolling-element bearing in which the rollers are small-diameter needle-like elements. Often used as a ***crankpin end bearing*** in small two-stroke engines.

needle valve (1) Valve by which fluid flow rate is controlled by the degree of insertion of a tapered needle into a fixed-diameter orifice. This type of valve can provide a fine gradation of flow. (2) Valve in which a blunt conically ended needle seats on a fixed-diameter orifice. With this type of valve there is little or no gradation of flow, control being limited to full, with needle lifted, or none, with needle seated. See Figure C.3.

negative camber Wheel camber in which wheel slopes inward toward the top.

negative offset steering Steering/suspension geometry in which point of intersection of ***kingpin axis*** with ground lies outside the center of the ***tire contact patch*** or plane of the wheel. See also ***center point steering***.

neoprene Chloroprene polymer ***synthetic rubber***, with a range of mechanical properties, noted for its resistance to oil and solvents at moderate temperatures. See also ***elastomer***.

net contact area Area enclosing pattern of the ***tire tread footprint***, excluding the area of grooves or other depressions.

net power Brake power of a fully equipped engine.

neutral Transmission idling position when no gear is engaged.

neutral start switch See ***ignition inhibitor***.

Neutralization Number Indication of acidity or alkalinity of an oil or fuel.

neutral steer (1) Having neither oversteer nor understeer. (2) At a given trim, having the ratio of steering wheel angle gradient to the overall steering ratio equal to the Ackermann steer angle gradient.

neutral turn Turn executed by a tracklaying vehicle in which one track only is driven, and the drive to the second track is in neutral.

Newton automatic clutch An automatic ***centrifugal clutch*** in which the centrifugal force on hinged weights overcomes a spring load to effect engagement. No clutch pedal is required, as engagement is a function of speed of rotation.

NiCad battery See ***nickel-cadmium battery***.

nickel A metal of good corrosion resistance and relatively high strength at elevated temperatures. Because of its high cost, it is used only in automobile applications such as chromium plating, and in the manufacture of high-grade magnets. As a structural material its application is mainly limited, as an alloy, to high-temperature components. Chemical symbol Ni.

nickel alkaline battery Storage battery with nickel hydrate positive plates and an alkaline electrolyte.

nickel-cadmium battery A nickel alkaline battery, of which the negative electrode is usually of a bonded cadmium and iron and the positive electrode of sintered nickel or nickel hydrate, with a dilute potassium and lithium hydroxide electrolyte. Used for traction and engine starting. Also ***NiCad battery***.

nickel-iron battery A nickel alkaline battery, of which the positive electrode is nickel oxide or nickel hydrate, and the negative electrode iron, with a potassium hydroxide electrolyte. Used for traction, but relatively heavier than lead-acid. Also ***nife battery*** from chemical symbols NiFe. (Informal) See also ***nickel alkaline battery***.

nickel-metal hydride battery Battery with a nickel or nickel-based positive electrode and a metal hydride negative electrode, capable of rapid recharge, and with high energy density and relatively low volume.

nickel-zinc battery Battery with positive electrode of nickel oxide and negative electrode of zinc, with a potassium hydroxide electrolyte.

Niemann Test A gear oil test.

nife battery See ***nickel-iron battery***.

nip Gap at mid-point between adjacent leaves of a loosely stacked ***leaf spring***.

nitrile rubber Synthetic rubber ***copolymer*** of butadiene and acrylonitrile, noted for its oil, fuel and temperature resistance, and therefore often used in shaft seals.

nitro Nitro-methane. (Informal)

nitro-methane Organic liquid, sometimes used as a fuel, with the addition of methyl-alcohol, in sprint competitions.

nitrobenzene An organic fuel ***additive***, highly poisonous.

nitrogen oxides Compounds of nitrogen and oxygen produced during combustion and conforming to the general formula NO_x.

non-dispersive infrared Analytical test method used mainly to determine carbon monoxide, carbon dioxide, nitric oxide, and hydrocarbon content of exhaust gases. Also ***NDIR***. See also ***infrared gas analyser***.

non-floating axle ***Live axle*** in which the axle shaft carries all chassis and torsional driving loads. See also ***fully floating axle***; ***semi-floating axle***.

non-powered axle Axle to which no power is transmitted. A ***dead axle***.

non-reactive suspension Tandem or multi-axle suspension in which driving and braking torques and axle loads on bump are equalized by mechanical linkage or other form of interaction. Usage of this term is confused (and confusing). It is sometimes taken to imply a ***tandem axle suspension*** where there is no equalization or compensation. See also ***reactive suspension***.

normal control (US: cab behind engine) Vehicle configuration, particularly of commercial vehicles, in which the engine is located forward of the cab, and is usually separated from the cab by a transverse ***bulkhead*** or ***firewall***. See also ***forward control***.

normally aspirated Unsupercharged. Breathing air at ambient pressure. Also ***naturally aspirated***.

normalized tire force coefficient See ***cornering stiffness coefficient***.

notchback Passenger car in which a sloping rear screen terminates in a short nominally horizontal extension.

NO_x See ***nitrogen oxides***.

nucleator Component of a cold flow improver additive, as of a diesel fuel, which creates nuclei onto which ***wax*** molecules attach themselves.

number plate illuminating lamp (US: license plate lamp) Lamp to illuminate registration plate or license plate. Also ***registration plate lamp***.

nut and lever steering See ***worm and nut steering***.

O

octane number Measure of the ***anti-knock*** properties of a fuel, particularly gasoline, derived from comparative tests using a standard variable compression engine. The anti-knock properties of the fuel under test are compared with those of a fuel comprising iso-octane (given an arbitrary rating of 100) and heptane (rated at zero). The volumetric percentage of iso-octane is taken as the octane rating of the fuel under test. Three test conditions are in use, the research method (giving ***Research Octane Number*** or ***RON***), the motor method, and the front-end method. American pump ratings are based on an average of ***Research*** and ***Motor Octane Numbers*** whereas European pump ratings quote ***Research Octane Number*** (which gives a higher value for comparable anti-knock properties).

odometer (UK: mileometer) Instrument that displays distance traveled by a vehicle.

off-highway vehicle Vehicle, as used for construction or agriculture, intended mainly for operation on unmetalled surfaces. Such vehicles may be permitted to use public highways.

offset Lateral distance at ***ground plane*** between point of intersection of ***steering axis*** and ***wheel plane***. See Figure C.2.

offset power steering Power-assisted steering system in which the powered steering effort is applied to the ***drop arm*** or ***pitman arm*** at the ***steering box***. See also ***in-line power steering***; ***linkage power steering***.

offset wheel Wheel with substantial distance between centerline of tire and outer ***nave face***, and particularly when nave face lies beyond the extreme width of the tire's sidewall, as on a twin wheel configuration of a heavy vehicle.

offside Side of a vehicle furthest from the kerb (curb) in normal driving. Generally the driver's side.

OHC See ***overhead camshaft***.

OHV See ***overhead valve***.

oil bath lubrication Lubrication by intermittent or continuous immersion in a bath of oil, or by an oil disc or ***flinger***.

oil control ring Piston ring which removes excess oil from cylinder walls of an engine and returns it via oilways in the piston to the main flow of oil. Usually the farthest from the ***crown*** of the piston rings. Also called ***oil ring***, ***oil-scraper ring***, ***and scraper ring***. (Obsolescent) See Figure P.2.

oil cooler Device to prevent overheating of engine or gearbox lubricating oil by natural or forced air convection or radiation, often on the principle of an engine cooling radiator.

oil dilution Reduction of lubricating oil viscosity, particularly of an engine oil by contamination with fuel residues.

oil disc See ***oil flinger***.

oil engine Compression ignition engine, so called because of the oily (viscid) nature of its fuel. (Obsolete term)

oil filter Device which removes suspended particulate matter from lubricating or hydraulic oil. Most engine oil filters filter by means of a paper, synthetic or ceramic element that traps particles in minute interstices. See also ***centrifugal filter***.

oil flinger Rotating disc the periphery of which is immersed in lubricating oil, which is spread by centrifugal action and conducted by oilways to bearings or other surfaces requiring lubrication. A now obsolete method of engine lubrication, also called oil-disc lubrication. Sometimes called ***oil slinger*** or ***oil disc***.

oil gage (1) **(UK: dipstick)** Graduated rod to indicate oil level in engine oil pan (sump) or transmission. (2) An oil pressure meter.

oil pan (UK: oil sump; sump) Bath-shaped reservoir fitted beneath the crankcase or cylinder block of an engine, and forming a container to which the lubricating oil returns by gravity. See Figure E.1.

oil pump Pump for distributing lubricating oil under pressure to the bearings and other lubricated surfaces of an engine or other machinery.

oil ring See ***oil control ring***.

oil-scraper ring See ***oil control ring***.

oil slinger (1) Flange or collar fitted to a shaft to disperse excess oil by centrifugal force. (2) An ***oil flinger*** or ***disc***.

oil sump (US: oil pan) Bath-shaped reservoir fitted beneath the crankcase or cylinder block of an engine, and forming a container to which the lubricating oil returns by gravity. Usually contracted to sump. See Figure E.1.

oil thrower See ***oil slinger***.

oilway (US: oil passage) Passage or conduit for lubricating oil under pressure, as in engine or other mechanism.

omnibus Public service vehicle designed to carry fare-paying passengers. Also ***bus***; ***motor bus***. See also ***trolleybus***.

on-center handling Ability of a vehicle to drive in a straight path without the need for continuous driver correction. The quality of straight-line handling of a vehicle.

once-through lubrication Lubrication system in which the lubricant is not recycled.

one-way clutch Mechanical clutch, such as a ***sprag clutch***, that engages when torque is applied in one direction of rotation only. When torque is not applied the driven member rotates freely. An ***overrunning clutch*** or ***freewheel***.

open cycle turbine Gas turbine operating cycle in which air enters from atmosphere and (with combustion products) is discharged to atmosphere. See also ***closed cycle turbine***.

open-loop engine control Control of engine parameters such as ignition timing and air-fuel mixture by pre-set, often mechanical or pneumatic, control such as ***centrifugal ignition advance***.

open-loop mode See ***position mode***.

opposed cylinder engine Engine in which cylinders are disposed at opposite sides of the crankshaft, usually horizontally. Informally designated by the number of cylinders as flat twin, flat four, etc. Also called ***boxer engine***. (Informal) See also ***flat engine***.

opposed piston engine Reciprocating engine in which pairs of pistons operate in each cylinder, the combustion chamber being formed between the piston crowns. Rare in current automotive use. See also ***deltic engine***.

opposite lock Steering input in opposite direction to current sense of turn, as in correcting a rear wheel skid.

optoelectronic materials Mainly polymeric materials that fluoresce when electrically charged, or develop an electric potential when exposed to luminous radiation.

organic compound Substance containing carbon and hydrogen, possibly with other elements. Not all organic compounds are necessarily of natural derivation, like coal and natural gas. Some are synthetically produced.

O-ring A toroidal elastomeric static seal.

oscillating warning lamp Lamp in which the beam or beams oscillate through fixed angles.

oscillation angle Of a ***beam axle*** suspension, the angle by which it can rotate about a longitudinal axis. Particularly applied to off-highway vehicles such as agricultural tractors and earth-moving machinery as a measure of ability to steer or provide traction on rough terrain.

Otto cycle The four-stroke spark ignition engine cycle, patented 1876, comprising induction (intake), compression, expansion (power), and exhaust strokes, the fuel

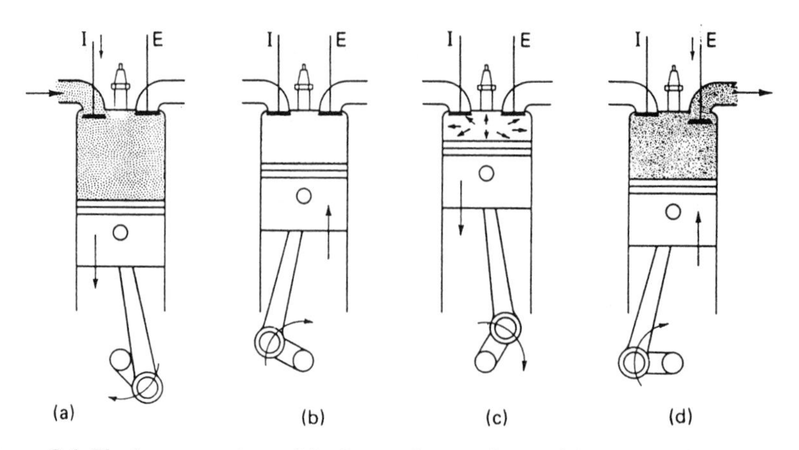

Figure O.1 The four operations of the Otto or four-stroke spark ignition cycle: (a) induction, (b) compression, (c) ignition followed by the working stroke, and (d) exhaust. I = inlet valve; E = exhaust valve.

and air mixture being compressed before combustion is initiated by an electrical spark or other means. After inventor N.A. Otto. See Figure O.1.

outboard brakes Brakes located at or near the wheel ***hub***. See also ***inboard brakes***.

outer dead center The equivalent of ***top dead center*** in a (horizontally or vertically) opposed engine. Infrequently used as equivalent to TDC of an in-line or single cylinder engine.

output ballast resistor See ***ballast resistor***.

outset dish wheel Wheel of which the ***nave face*** is inboard of the tire centerline.

oval piston Engine piston manufactured to slightly elliptical cross section to counteract the effects of thermal distortion. See also ***cam-ground piston***.

overall steering ratio Rate of change of steering wheel angle to that of steered wheels, at a particular angle and assuming no play or mechanical losses in the system.

overdrive An auxiliary ***gearbox***, usually of ***epicyclic*** or ***planetary*** type, located between main gearbox and rear axle, and employed to reduce engine speeds for economical high-speed cruising. The overdrive can usually be switched into and out of operation by electrical or mechanical control.

overdrive gear Gearcase/gearbox with a ratio of less than unity.

overhang (1) Longitudinal distance that a vehicle extends beyond each wheel axis. See Figure L.3. (2) Longitudinal distance that the load of a goods vehicle extends beyond the end of the vehicle, but subject to local or national definition, particularly for multi-axle vehicles, where an intermediate point between centerlines of rear axles is taken as datum. Also ***load overhang***. See also ***approach angle***; ***departure angle***; ***ground-line angles***.

overhead camshaft Of an engine, having the ***camshaft*** located in the cylinder head and operating the valves directly or indirectly from above. See Figure O.2.

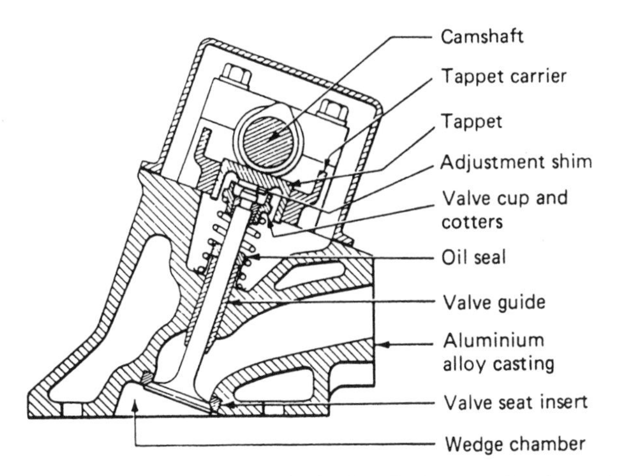

Figure O.2 Section through cylinder head of an overhead camshaft engine with a wedge-shaped combustion chamber.

overhead valve Valve mounted and seated in the cylinder head of an engine. Also ***OHV***.

overhead valve engine (US: I-head) Engine having the valves located at the head of the combustion chamber, opposite to the piston, and operated by a mechanism of rockers and pushrods from a camshaft mounted in the cylinder block. See also ***side valve engine***; ***overhead camshaft***. See Figure O.3.

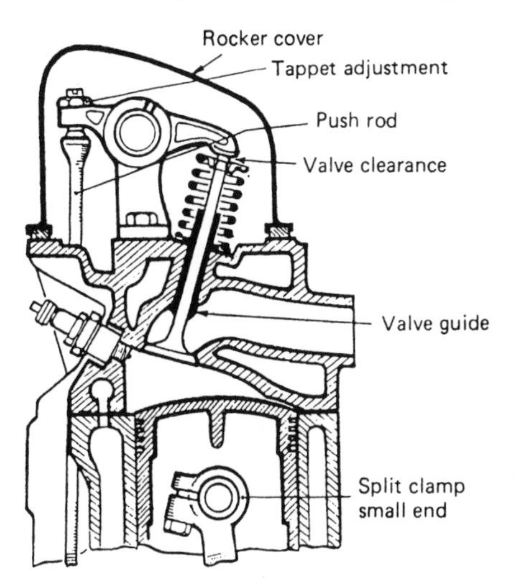

Figure O.3 Section through cylinder head of an overhead valve or I-head engine.

overlap See ***valve overlap***.

overlay To locally face with a hard or wear-resistant material, as of a cam. (Mainly US usage)

overrider (US: bumper guard) Short vertical bar attached, usually on either side, to a vehicle bumper to prevent interlocking of bumpers in minor collisions. Also ***mae west***. (US slang)

overrun Motoring of an engine by way of the inertia of a vehicle in motion, with throttle control closed.

overrun braking See ***inertia braking***.

overrun fuel cutoff Valve, usually electronically controlled, that inhibits fuel supply on overrun.

overrunning clutch Mechanical device that disengages the engine on ***overrun***, and so acts as a freewheel.

overrunning clutch starter motor Starter motor in which engagement of pinion with ring gear is brought about by lever action from a solenoid or by manual operation. An ***overrunning clutch*** or ***freewheel*** prevents the driving of the pinion before disengagement. See also ***Bendix drive***; ***pre-engaged starter***.

overslung worm transmission Worm and worm wheel final drive in which the worm is set above the worm wheel. Used in commercial vehicles and where ground clearance is important, though now obsolescent. See also ***underslung worm transmission***.

overspeed governor Device that limits the speed of an engine, usually by controlling the fuel supply.

overspeeding Of an engine, exceeding design speed, or a speed limited in accordance with emissions, noise or other regulations.

oversquare engine Engine having a cylinder bore larger than its stroke. See also ***square engine***; ***undersquare engine***.

oversteer (1) Response of a vehicle if the ratio of steering wheel angle gradient to overall steering ratio is less than the Ackermann steer angle gradient. (2) Over-response to steering input, as by generation of excessive ***slip angle*** on rear wheels. (3) Response of vehicle to steering input characterized by an incremental increase in ***yaw rate*** which necessitates a decrease in steer angle to maintain the intended radius of turn. See Figure O.4.

ozone cracking Failure mode of elastomers when under cyclic strain in the presence of ozone. Failure can occur at below the mechanical fatigue limit of the material.

ozone resistance Resistance of ***elastomers*** and ***polymers*** to attack by oxone under static and dynamic conditions.

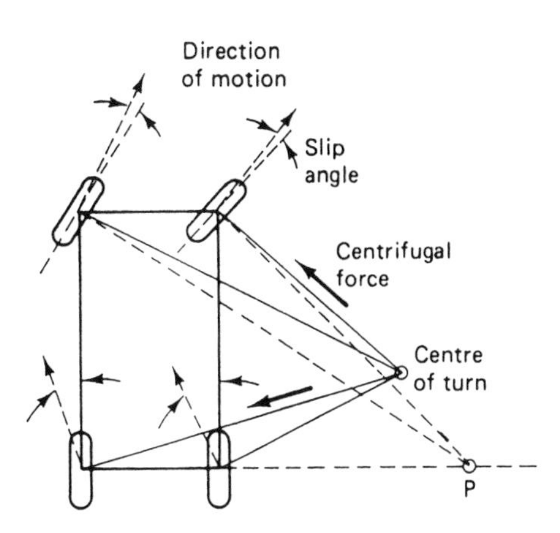

Figure O.4 Diagrammatic representation of oversteer, showing the actual center of turn compared with the geometric center of turn that would prevail at low speed.

oxygenates Alcohol-based ***fuels*** or fuel ***additives***.

P

packing See *gasket*.

pad The friction lining of a ***disc brake***.

pallet truck Industrial vehicle equipped with means of lifting and transporting freight set on pallets. See also ***forklift truck***.

palm coupling Combined mechanical and pneumatic coupling for attaching a trailer and its ***air brake*** system to a towing vehicle.

panel Sheet metal component of a vehicle body, particularly when part of the outer shell.

panel body Fully enclosed truck or van body, constructed from panels of sheet metal or other material. See Figure L.5.

Panhard rod Transverse rod pivoted at one end to chassis or vehicle shell and at the other to a ***beam axle***, which it constrains in lateral movement.

pannier (1) Originally bags or baskets carried either side of a pack animal. In automotive context describes specially fashioned bags or boxes set on either side of a bicycle or motorcycle wheel. (2) External tanks carried on either side of vehicle.

pantechnicon (US: moving van) Large-capacity van, particularly one equipped for the transportation of bulky goods or household effects. (UK obsolete)

parabolic spring Leaf spring of parabolic taper profile, typically consisting of one to four leaves, and of lower stack height than an equivalent multi-leaf spring. See also ***semi-elliptical spring***.

paraffinic oils Lubricating oils containing a high proportion of ***paraffin*** molecules, with generally high ***pour points*** but good viscosity/temperature characteristics. Lower in solvency than napthenic oils.

parasitic drag Aerodynamic drag caused by the projection of objects into the airflow such as radio antennae and door handles. Also ***parasite drag***.

park (1) Condition of automatic transmission vehicle, with or without engine running, transmission disengaged and drive wheels locked. (2) To find a stopping place where a vehicle can be left unattended.

parking brake Brake system that holds one or more brakes, or the transmission in an automatic transmission vehicle, permanently on when vehicle is parked. Also ***handbrake*** in a manual transmission vehicle.

parking lamp Lamp to show the presence of a parked vehicle, white to front and red to the rear. In some systems the offside lamps only may be activated. Also ***side-light***. (Non-preferred)

parking torque Turning effort required in steering a vehicle at very low speed.

part-time case Four-wheel-drive transfer gearcase that permits disengagement of drive to one axle.

partial skirt piston See ***slipper skirt piston***.

particulate filter A filter to remove air-suspended particulates, as in the exhaust of a diesel engine.

particulates Solid particle content of the exhaust products, mainly in the form of carbon (soot) and partially burned hydrocarbons. See also ***emission***.

passenger car Vehicle with motive power designed for carrying usually ten persons or less, and not for hire or reward.

passenger compartment Part of a vehicle body that houses the passengers and driver, except where the driver occupies a separate cab.

passive restraint Device to restrain vehicle occupant in event of an impact, and which does not require any action of the occupant, such as fastening or adjustment.

passive safety Safety qualities of a vehicle assuming an impact has taken place, for example structural crashworthiness, energy absorption of steering column, etc.

passenger vehicle Vehicle constructed for carriage of passengers and their effects. See also ***bus***; ***coach***.

patch (1) Area of contact of ***tire*** with ground. (2) Repair to an inner tube.

pay cube Volume of the load-carrying body of a commercial vehicle that is capable of carrying payload.

payload The cargo of a goods vehicle, or the weight of the cargo, as subject to local definition: effectively, the ***gross weight*** less the ***tare weight*** under UK legislative definition.

PCV See ***positive crankcase ventilation***.

peak brake power Highest power developed within speed range of an engine. Most commercial diesel engines are governed to operate below peak brake power speeds.

peak power engine speed Engine speed at which peak power occurs. See also ***brake power***.

peak torque speed Engine speed at which peak torque occurs. Not necessarily the same as peak power speed.

peaky Of an engine, developing effective power over a narrow speed range, thus requiring frequent gear shifting under varying load. See also the antonym ***flexible***. (Informal)

pedal Foot-operated control.

pedal effort Load applied by the driver's foot to depress a pedal.

peel test Test for strength of adhesion between two materials, particularly where at least one of the materials being joined is a flexible lamina.

pendulous absorber See ***pendulum damper***.

pendulum damper Device employed particularly in diesel engines to attenuate crankshaft vibration by vibrating in antiphase. Also ***pendulous absorber***.

pentane Light and volatile paraffinic constituent of ***gasoline*** which aids engine starting.

pentroof Shallow angled wedge or cone, particularly of a ***piston crown***. Often used to describe the shape of a four-valve combustion chamber head.

Perbury drive A design of toric infinitely variable transmission.

perch bracket The mounting bracket for a center-mounted ***beam steering axle***, as of an agricultural tractor or steam traction engine.

permeability Measure of the ability of a liquid, vapor or gas to pass through a solid material such as an ***elastomer*** or natural rubber.

petrochemical Any chemical substance, such as a fuel, oil or plastic, derived from crude oil or its products, or from natural gas.

petroil Gasoline mixed with lubricating oil for use in a two-stroke engine. (UK informal)

petrol (US: gasoline) Volatile hydrocarbon liquid fuel for spark ignition engines. Also ***petroleum spirit***; ***motor spirit***. See also ***octane number***.

petrol engine (US: gasoline engine) Spark ignition engine using gasoline (petroleum spirit) as a fuel.

petrol tanker (US: gas truck) Heavy vehicle equipped for conveying and delivering petrol (gasoline).

phenolics Thermosetting plastics of many forms and based on phenolic resins, to which are usually added fillers and colorants. Phenolics are generally strong and dimensionally stable, with good electrical insulating properties. One of the earliest automotive plastics.

phosphor bronze A bronze containing small quantities of phosphorus, and noted for its bearing properties, which may be enhanced by the addition of ***lead***.

pickup See ***pick-up truck***.

pickup coil Electrical speed sensor for ***electronic ignition***.

pickup lag Delay of response of an engine to ***accelerator*** pedal depression.

pick-up truck Light capacity open truck, usually with a lowering tail gate. See also ***ute***. Also ***pickup***. (Informal)

pigment Colored particulate for imparting color to paints, finishes, and plastic moldings and laminates.

piling See ***pitting and piling***.

pilot jet Carburetor ***jet*** that admits fuel downstream of the throttle for starting.

pilot spray Spray that initiates combustion and precedes the main spray in some types of diesel injection system.

pin synchronizer Transmission synchronizer in which the actuating hub is loosely pinned between two stop rings.

pin boss See ***piston pin boss***.

pin slide caliper A sliding or floating caliper ***disc brake***.

ping (UK: pinking) ***Detonation*** or partial detonation in engine cylinder, or the noise thereof. (US informal) See also ***knock***.

pinion A small gear wheel, or the smaller of a meshing pair of gear wheels.

pintaux nozzle Diesel ***fuel injector*** of ***pintle*** type with auxiliary hole through which a fraction of the fuel is directed to the combustion chamber center. Mainly used in ***Comet*** type ***indirect injection*** systems.

pintle (1) Vertical pin or spigot to which a tow bar is attached. Also pintle hook. (2) Gas flow control valve. See also ***pintle nozzle***.

pintle nozzle Diesel fuel ***injector nozzle*** in which a spring-loaded needle valve (or pintle) controls a hollow cone-shaped fuel spray. See Figure P.1.

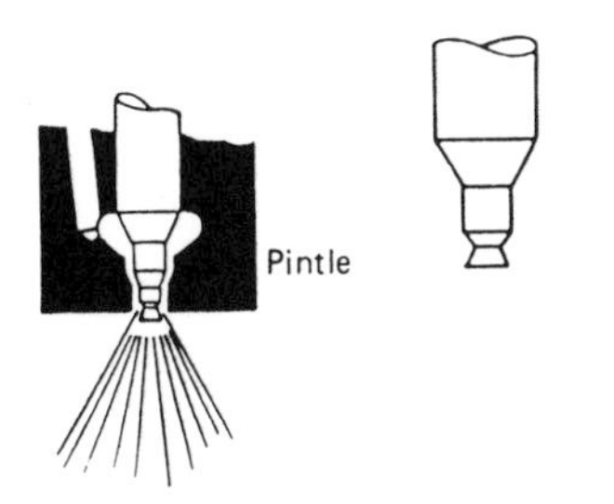

Figure P.1 Pintle nozzle of a diesel injector.

pipe A tubular conduit, usually rigid. See also ***hose***; ***tube***.

piston Component, usually in the form of a cylinder closed at one end, that converts fluid pressure into mechanical movement and force, or *vice versa*, within a smooth walled cylinder, in which it is a sliding fit. In a reciprocating engine gas pressure on the ***piston crown*** provides the prime force that is converted into rotating mechanical power by the ***crankshaft***. Pistons are found in hydraulic and pneumatic systems. See also ***disc brake***. See Figure P.2.

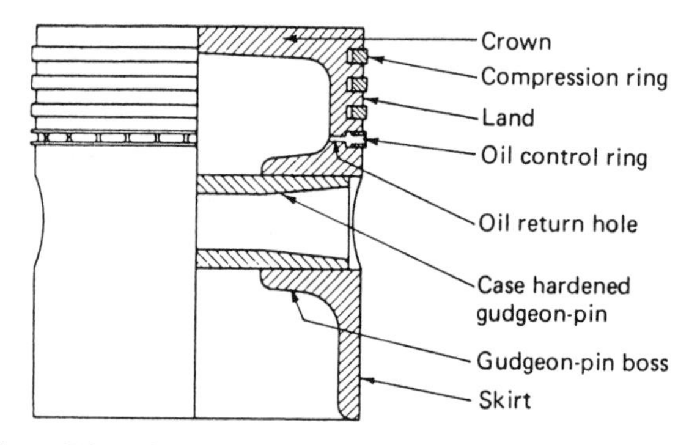

Figure P.2 Half-section of a piston.

piston boss Sturdy reinforced bearing area within an engine piston that is bored to take the ***piston pin*** (***gudgeon pin***).

piston cap False or additional crown, often of different material to a piston, that is fitted to a piston, usually to improve durability. See also ***piston crown***.

piston crown Closing end of a piston, of various shapes depending on engine type and valve configuration.

piston pin (UK: gudgeon pin) Pin that connects the piston to the ***connecting rod***. Also ***wrist pin***. (US)

piston pin boss Strengthened part of piston which holds the ***piston pin*** or ***gudgeon pin***. Also ***pin boss***. (US)

piston ring Ring, generally of hard, springy material, set in groove beneath ***piston crown*** or in ***piston skirt***. Internal combustion engines normally have three to five rings, each fulfilling a particular function. See also ***compression ring***; ***oil control ring***.

piston skirt Nominally parallel-sided cylindrical walls of a piston, extending beneath the ***piston crown***. See Figure P.2.

piston slap Noise made by contact between a loose or worn piston and the cylinder wall of an engine.

piston speed Linear velocity of a piston in its reciprocating motion within a cylinder. The term should be qualified as ***instantaneous piston speed*** or ***mean (average) piston speed***.

Pitman arm (UK: drop arm) Lever that converts rotary output from ***steering box*** to linear movement of a ***drag link***. Also ***sector shaft***. See Figure S.8.

pitting Chemical or mechanical surface damage involving the formation of pits or holes. See also ***scuffing***.

pitting and piling Erosion of one ignition contact breaker contact and deposition of metal on the other resulting from high temperature vaporization or incorrect condenser capacity.

pivot pin (1) A hinge pin about which part of a mechanism rotates or partially rotates. (2) Vertical or inclined shaft about which a steered wheel assembly pivots. Also ***swivel pin***; ***knuckle pin***. (US informal) Also ***kingpin***.

pivot ring Ring or raised annulus in a ***clutch casing*** that acts as a fulcrum for a ***diaphragm spring***. Also ***diaphragm spring ring***.

pivot steering Rotation of a (track-laying) vehicle without significant forward motion, as in pivoting about one track.

plain bearing A shaft bearing, usually of cylindrical form, in which a shaft rotates lubricated by an oil or grease, or, in some cases using the low friction qualities of a polymer such as PTFE or a solid lubricating agent in a porous material. See also ***main bearing***; ***big end bearing***; ***rolling element bearing***; ***shell bearing***; ***thin-wall bearing***.

planet carrier Disc or spider that carries the planet wheels in an ***epicyclic*** or ***planetary gearbox***. See Figure P.3.

planet wheel Spur gear wheel of an ***epicyclic gear*** train that meshes with the central sun wheel and outer internally toothed annulus. See Figure P.3.

planetary transmission (UK: epicyclic gearbox) An ***epicyclic*** gear system, consisting in basic form of an internally toothed ***annulus*** or ring gear and a central externally toothed ***sun wheel***, with usually three or four ***planet wheels*** or pinions meshing with the sun wheel and annulus. The planet wheels are often carried as a unit on a planet carrier or spider. Arresting the rotation of the ***planet carrier***, annulus or sun wheel will provide a transmission ratio between the two remaining elements. Planetary gearsets can be coupled to increase the available number of ratios. See Figure P.3. See also ***automatic transmission***; ***pre-selector***; ***Wilson gearbox***.

plasticiser Organic compound added to plastics and rubbers to make them more flexible.

plastisols Mainly heat-curing ***adhesives*** derived from PVC, and suitable for the assembly of larger automotive components, particularly where subjected to heat.

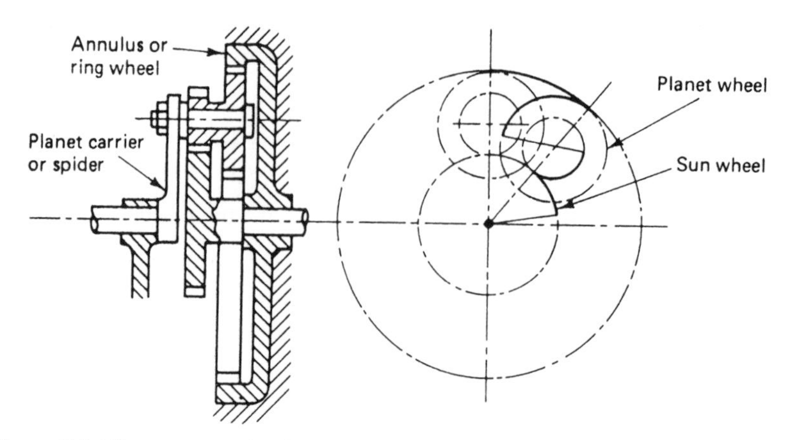

Figure P.3 Planetary or epicyclic transmission.

plate (1) Permanent notice attached to a commercial vehicle to indicate authorized conditions of use and loading. (2) A positive or negative plate of a battery. (3) A clutch plate.

plate load Force applied by clutch springs in engaging a clutch.

plated weight Maximum permitted gross weight of a vehicle or its individual axle loadings, as approved by current local or national legislation. See also ***plate***.

platform (1) Horizontal load-bearing surface or floor, as of a lorry, truck or van. See also ***loadfloor***. (2) Entry staging of a bus or coach.

platform body The body type of a platform lorry or truck. See also ***dropside***; ***float***; ***headboard***.

platform lorry (US: flatbed truck) Truck or lorry carrying a flat and generally flexible platform to which bulk goods or equipment may be lashed with cordage or belts.

platform trailer ***Trailer*** or ***semi-trailer*** with a load-carrying platform. See ***platform lorry***.

platinum catalyst Metallic platinum coating within exhaust gas catalytic converters that accelerates the oxidizing reaction of carbon and hydrocarbon based emissions and breaks down NO_X without itself undergoing any change. The effectiveness of the platinum is nullified by lead-based fuel ***additives***. Platinum is usually used in combination of other metals such as rhodium and palladium.

play Free movement or looseness within a mechanism, as for example any free movement of a steering wheel before the steered wheels respond by turning. Also ***lash***.

plenum chamber Any chamber at which air or gas is held at higher than ambient pressure.

plug (1) A (usually removable) stopper or closing element to prevent pressure loss or leakage. (2) Removable electrical connector, normally inserted into a fixed socket. (3) A ***spark plug***. (Informal)

plug lead Lead which conducts high voltage to spark plug. Also ***spark plug lead***.

plunger pump Reciprocating pump, with valve chest, usually for delivering liquids under pressure. Also called ***force pump***.

plunging joint A shaft joint or coupling with provision for axial or plunging movement. See also ***pot joint***.

ply Layer of rubberized fabric from which a tire ***carcass*** is made.

ply rating Index of strength of a ***tire***, originally but no longer the number of plies. Also ***PR***.

ply turn-up See ***flipper***.

plysteer Tendency of a loaded ***tire*** (as on a vehicle in motion) to pull left or right due to asymmetric effects of the plies. This will tend to pull a moving vehicle in one direction moving forward, and in the opposite direction moving in reverse.

plysteer force The force necessary to overcome the effects of ***plysteer***, and so keep the ***tire*** on a straight path on a flat, even surface.

PM10 Particulates of below ten microns in size, presenting specific health hazards.

pneumatic suspension Suspension using air or other gas in compression as a suspension medium. See also ***air suspension***; ***Hydragas suspension***.

pneumatic tire Hollow rubber toroid inflated by air pressure and mounted on the rim of a ***wheel***. Some definitions stipulate that a pneumatic tire must collapse if not inflated when subject to normal vehicle loads, and can be inflated without removing it from the wheel or vehicle. Also ***tyre***. (UK) See also ***cover***.

pneumatic trail (1) Distance that the point of effort of the total lateral force or cornering force on a ***tire*** lies behind the center of area of the ***tire contact patch***. Numerically equal to self-aligning torque divided by lateral force. (2) Horizontal distance between point of action of the side force due to the ***slip angle***, or the ***camber thrust***, and the ***tire contact point***.

pocket Relieved flank of a Wankel engine's rotor, forming part of the combustion chamber.

point The switching contact of an ignition ***contact breaker***.

poisoning Of an exhaust ***catalyst***, degrading or rendering ineffective by action of lead or other compounds in exhaust gas.

pole trailer (1) ***Full trailer*** drawn by a pole or long circular section towbar or ***reach***, which is connected to a ***steering axle***. (2) ***Semi-trailer*** or ***full trailer*** in which a central tube serves as a chassis, used for transporting lumber or other long and relatively rigid items. See also ***spine back***.

pollutant Substance released into the environment which represents a hazard to man or nature.

polyaryletherketone A high-temperature molding ***polymer***.

polycarbonate Range or thermoplastic and thermosetting resins of high toughness over a wide temperature range, good electrical properties, and self-extinguishing. Used for injection and blow moldings, extrusions and vacuum forming.

polyester Range of thermosetting plastics or resins used in the making of GRP bodies and parts, and noted for toughness and weather resistance.

polyglycols Polymers of ethylene and/or propylene oxides, used as ***synthetic lubricants*** and ***hydraulic fluids***.

polyisoprene Synthetic rubber, with similar properties to natural rubber, made by the polymerisation of isoprene.

polymer Substance made by linking together molecules of similar or different types (called ***monomers)*** to form substances of high molecular weight.

polyolester Synthetic ***lubricant*** made by reacting fatty acids with polyhydric alcohols.

polysulfone A high-temperature molding polymer.

poppet valve Disc-shaped valve, the disc being attached to a ***stem***, the reciprocating movement of which causes the valve to open and close. The valve disc edge is chamfered to facilitate sealing, location and heat transfer. Also called ***mushroom valve*** because of its shape. The normal type of valve in an internal combustion engine. See Figure E.1.

popping back Premature ignition of a fuel/air mixture in an ***inlet manifold***.

port Hole or aperture shaped to facilitate the flow of gas or liquid into or from a chamber. Usually denoted by its function, as ***inlet port***, ***exhaust port***.

port closing Of a ***fuel injection pump***, the point of closing of the port by the metering member, corresponding to the nominal start of pump delivery.

port opening Of a ***fuel injection pump***, the point of opening of the port by the metering member, corresponding to the nominal end of pump delivery.

position mode Engine test procedure carried out at fixed throttle opening and fixed dynamometer setting. Also ***open-loop mode***.

position and power-law mode Engine test procedure at fixed throttle opening but with dynamometer setting varied to produce a stipulated torque/speed characteristic.

position and speed mode Engine test procedure from which torque/speed curves are derived.

positive camber Wheel camber in which the wheel slopes outward towards the top. See Figure C.2.

positive caster Steering geometry in which the ***center of tire contact*** lies behind the point at which the ***steering axis*** intersects the road surface, thus giving a positive or stable self-centering effect. Also ***trailing caster***. See Figure C.2.

positive crankcase ventilation Emission control system that prevents crankcase gases from entering the atmosphere, usually by drawing the gases from the ***crankcase*** and feeding them into the engine's ***induction system***. Also ***PCV***.

positive displacement pump Pump that delivers a fixed or metered amount of gas or liquid per stroke or cycle.

positive offset steering Steering/suspension geometry in which point of intersection of ***kingpin axis*** with the ***ground plane*** lies inboard of the center of the ***tire contact patch*** or plane of the wheel. See also ***center point steering***.

post (1) Nominally upright structural member, particularly for supporting the roof and doors of a vehicle. (2) A battery terminal. (US informal)

post-ignition Ignition emanating from ***hot spots*** within an engine cylinder and occurring after the main spark-initiated ignition.

pot flywheel Engine flywheel with extending flange or outer rim, usually to accommodate ***clutch*** assembly.

pot joint Universal joint in which the crosspin rollers are allowed axial movement in an internally slotted cylinder. A ***plunging joint***. Also ***de Dion*** or sliding-block joint. See also ***tri-pot joint***. See Figure P.4.

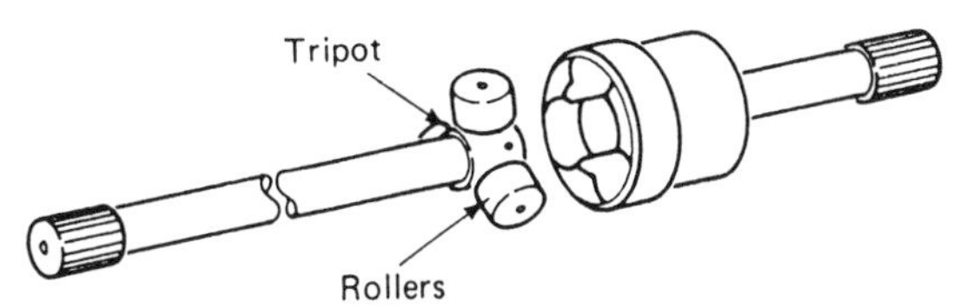

Figure P.4 A tri-pot type of pot or plunging joint.

pour point (1) Of a fuel oil, nominally the temperature three degrees Celsius above which the fuel will just flow under its own weight. A measure of the low-temperature flow characteristics of a fuel. (2) Of a lubricant, a test for the gelling tendency of paraffinic oils at low temperatures.

power-assisted steering Steering in which driver effort is aided by mechanical, hydraulic or other power, while retaining some degree of road interaction feedback.

power brakes Brakes operated by mechanical, hydraulic or pneumatic power rather than by driver effort.

power density Of an electrical battery, the power available per unit volume, usually expressed in Watts per litre. See also ***energy density***.

power divider Geared unit that splits the power from one input shaft between two or more axles of a ***tandem axle*** arrangement. See also ***transfer box***; ***transfer case***.

power hydraulic system In context of braking, a system in which hydraulic fluid is stored under pressure and is metered out to the brakes by driver operation of a valve actuated by the ***brake pedal***.

power jackknifing Jackknifing of an articulated vehicle resulting from spin or slip of the wheels under high driving torque, as for example when accelerating or climbing a gradient. See Figure J.1.

power output Power developed by an engine, normally qualified by the means of measurement or test specification, and other parameters affecting operation. See ***brake power***; ***indicated power***; ***peak brake power***.

power shift PTO Power take-off remotely controlled by mechanical, hydraulic or other form of power control.

power steering (1) Steering in which angular movement of the steering wheel actuates a powered steering system, there being no mechanical feedback between road and driver. (2) A power-assisted steering system. (Informal [and inaccurate]) See also ***in-line power steering***; ***linkage power steering***; ***offset power steering***. See Figure R.1.

power stroke Stroke of a reciprocating engine during which the piston moves under the effect of combustion pressure, normally taken from ***top dead center*** to ***bottom dead center***. Preferred term, but also called combustion, working or expansion stroke. See Figure O.1.

power take-off (1) Driven shaft with external coupling on engine or transmission system to which ancillary equipment may be attached, as for example agricultural machinery to a tractor. (2) Separate transmission unit for attachment to an engine, gearbox or part of a vehicle's transmission system, with external controls, for the driving of ancillary equipment, particularly when the vehicle is stationary. Also ***PTO***.

powered axle See ***live axle***.

powered speed Geared speed from which is subtracted losses resulting from actual impediments such as air resistance, slip, etc.

powertrain The components making up the power transmission system of a motor vehicle, from ***clutch*** to ***final drive***. In some instances the term may include the engine. See also ***drive line***. See Figure P.5.

PR See ***ply rating***.

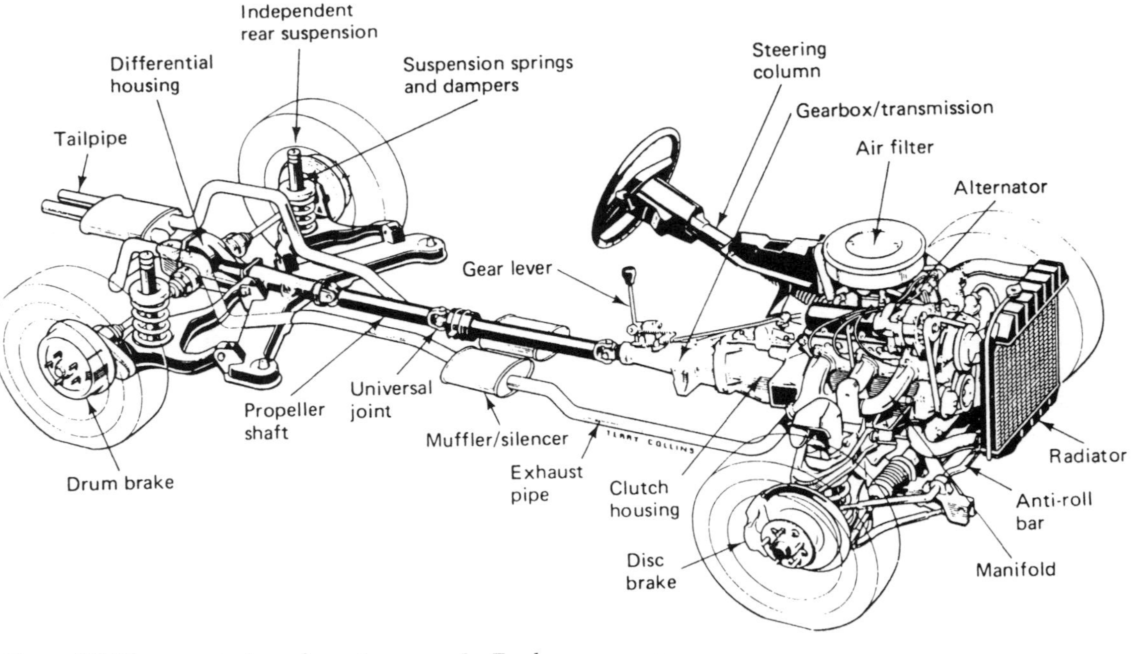

Figure P.5 The powertrain and running gear of a Ford passenger car.

prechamber Small chamber in which combustion is initiated prior to delivery by way of a narrow port into the main combustion chamber of an engine. Prechambers are of widely varying design and are generally formed in the ***cylinder head***. It is normally understood that fuel is injected within a prechamber, whereas it is not necessarily injected into an ***air cell***. The term ***indirect injection engine*** is used in the UK and ***precombustion engine*** in the US to describe engines employing prechambers. Also called ***antechamber***. See also ***Comet head***; ***Lanova air-cell***.

precombustion engine (UK: indirect injection engine) Engine, particularly a ***diesel engine***, employing a ***prechamber***.

precompression Compression of a gas prior to the main compression cycle.

pre-engaged starter Positive action solenoid-operated ***starter motor*** in which a pinion engages with a ***ring gear*** before the full electric current flows. The pinion is free to freewheel before disengagement when engine has started. See also ***overrunning clutch starter motor***.

pre-heated catalyst Exhaust ***catalytic converter*** that heats up on or before starting of the engine, and is not therefore delayed in operation by the need to absorb heat from the engine exhaust before becoming effective.

pre-ignition Ignition of the fuel-air mixture in a spark ignition engine before the timed spark, as caused by a hot or glowing surface or other agency. (Obsolete and technically misleading term) See also ***spark knock***.

pre-selector Semi-automatic ***change-speed gearbox*** in which the gear is selected before change is initiated, change being made by depressing a pedal. See also ***Wilson gearbox***.

pressure charge induction See ***harmonic induction engine***; ***ram air induction***.

pressure drag Aerodynamic drag resulting from difference between pressures acting on forward and rearward facing surfaces. See also ***friction drag***; ***induced drag***.

pressure plate Rigid disc-shaped ***clutch*** element that acts under spring load against the friction lining of a ***clutch driven plate*** (or clutch disc) to effect engagement. See Figure C.4.

pressure plate spring Spring, either a diaphragm spring or one of a group of coil springs, that holds the clutch ***pressure plate*** in contact with the ***clutch driven plate*** (or clutch disc). See Figure C.4.

pressure protection valve Pneumatic valve that prevents inadvertent loss of air from a pressurized system, as for example the ***air brake*** system of an articulated vehicle when the trailer and tractor are uncoupled.

pressure wave supercharging Increasing engine intake pressure by tuning the intake duct to resonate at a particular operating speed. Also ***ram-pipe supercharging***.

pressurized cooling system Sealed engine cooling system which uses the vapor pressure of the hot coolant to create an above ambient pressure in the system and thus raise the boiling point of the coolant.

primary In context of engine balancing, related directly to engine speed (rather than to twice engine speed). See also ***primary couple***; ***primary force***; ***secondary***.

primary couple Out-of-balance mechanical disturbance in an engine that causes rocking about a horizontal or transverse axis (in a vertical or V configuration engine) at right angles to the ***crankshaft***, and at engine speed. Also called pitching moment if about transverse axis and yawing moment if about vertical axis.

primary force Out-of-balance force (in the plane of the cylinder axis in an in-line engine, or its bisection in a V configuration engine) resulting from disparity of static mass balance of rotating and reciprocating parts. Also called ***free inertia force***.

primary reference fuel Combustible blend used as a standard for ***knock*** evaluations in spark ignition engines and for ***Cetane Number*** evaluations in diesel engines. Also ***PRF***.

primary shoe ***Leading shoe*** or a ***drum brake***. Also ***forward shoe***. (US) See Figure C.1.

priming Introducing fuel into the ***induction system*** of an engine prior to starting.

producer gas Predominantly hydrogen gas produced by passing air and steam over burning coke or coal. Formerly used as an engine fuel in times of acute shortage of conventional fuels. See also ***gas generator***.

production test cell Test cell for testing newly built engines and other equipment.

profile drag The aerodynamic drag of a vehicle attributable to the vehicle's external shape, and not to its surface texture or total surface area. See also ***drag***; ***streamlining***.

progressive spring Spring in which the spring rate increases with deflection.

propane Hydrocarbon gas fuel stored in liquid state under pressure. A constituent of ***liquefied petroleum gas***.

propeller shaft (US: drive shaft) Rotating shaft, usually tubular, that transmits the drive from the main change-speed gearbox to the differential or final drive. Also ***drive shaft***. See also ***Hotchkiss drive***. See Figure H.2.

proportional control Closed-loop engine control system in which the feedback signal is proportional to output of the measured parameter. See also ***limit cycle controller***.

proportioner Valve in a hydraulic or pneumatic system that apportions effort between two or more circuits, as in a split braking system. See also ***compensator***.

PTO See ***power take-off***.

public service vehicle Vehicle intended to carry passengers (usually twelve or more, not including the driver) for hire or reward. This definition is subject to local variation, particularly regarding the minimum number of passengers. Also ***PSV***. See also ***omnibus***.

public works vehicle Mechanically propelled vehicle for use by or on behalf of a public authority or other statutory body as defined by Local Government Act legislation in the UK. The definition includes post office, police and telecommunications vehicles.

pull-off spring Spring to assist the return to inactive position of a mechanically or hydraulically actuated item, for example the return to the non-applied position of brake shoes in a drum brake.

pull-type clutch Clutch with a clutch release or throwout mechanism operated by a pull action, as used on heavy vehicles where depression of the clutch pedal also operates a gearbox brake to quicken the operation of ***synchronization***. See also ***push-type clutch***.

pullrod Any rod that transmits linear displacement to an actuating mechanism when in tension.

pulse air injection System that uses exhaust pressure pulsations to force air into an exhaust system to oxidize exhaust products. See also ***air check valve***.

pulse wheel Indexed or castellated wheel whereby timed electrical pulses are sent to the spark plugs of an engine ignition system, usually replacing the conventional ***distributor***.

pump Mechanical device which causes liquids, gases or vapors to flow by means of pressure differential or positive displacement. See ***oil pump***; ***water pump***.

pump control rack Rack engaging with pinion that sets position of fuel delivery plunger in ***fuel injection pump***.

pumping losses Part of the total or indicated power of an engine that is expended on the induction of the fuel and air charge into an engine, and expelling the gases on completion of combustion. See also ***brake power***.

puncture A hole in a tire through which pressurized air escapes. See also ***blow-out***.

punt chassis Chassis consisting of a continuous box-structure, with side flanges and central well, to which a nominally non-structural body is attached.

purge (1) To drain and remove all traces of a fluid from a container or system. (2) In exhaust emission testing, to remove residual gases from test equipment.

purge cock (UK: drainplug, drain tap) Outlet to facilitate draining or emptying a working fluid or lubricant from an engine, gearbox, etc.

push rod See ***pushrod***.

push-type clutch Clutch with a clutch release mechanism operated by a push action. See also ***pull-type clutch***. See Figure C.4.

pusher axle (1) Rear-powered axle of a ***tandem axle*** arrangement, when the forward axle or axles are non-powered or "dead." (2) Non-powered (dead) axle of tandem axle arrangement, located ahead of the ***drive axle***. (US usage)

pusher tandem ***Tandem axle*** arrangement in which only the rear axle is powered.

pushrod Rod that transmits linear displacement to an actuating mechanism when in compression, as for example ***cam lift*** of a four-stroke reciprocating engine to the valve gear. See ***pushrod engine***. See Figures C.14 and V.1.

pushrod engine Reciprocating engine in which the valve mechanism is actuated by intermediate ***pushrods*** rather than directly by the cam (as in an ***overhead camshaft*** engine). Usually describes an ***overhead valve*** engine in which the pushrod transmits the linear displacement or ***lift*** of the cam to a rocker lever, the camshaft normally being at a lower level than the cylinder head. (Informal) See Figure O.3.

pyrene One of a group of complex ***hydrocarbon*** products of combustion (called polynuclear aromatic compounds) which are of high toxicity and carcinogenicity. A constituent part of the unburned hydrocarbon content of exhaust gases. See ***hydrocarbon***.

Q

quadrant (1) The toothed sector used in a ***worm and sector*** (or worm and quadrant) steering gear. (2) Position indicator of an automatic transmission gearshift lever. (3) Any mechanical item of quadrant form.

quadrant change Gear lever with one plane of operation and gear selection sequential in that plane. Obsolete in manual shift/change systems.

qualifying tire Racing tire of high frictional coefficient but short life for achieving fast times on practice laps to qualify for favorable starting grid position.

quarter elliptic Cantilevered half of a ***semi-elliptic leaf spring***. Not, however, the same as a cantilever spring.

quarter light (1) Rear side window of a passenger car or truck cab, particularly when set aft of the door. (2) Vertically narrow side window set aft of the opening window of a rear door, which may be vertically hinged for opening.

quarter wavelength resonator Acoustic device of cylindrical form with a pre-set or adjustable resonant frequency determined by its length. See also ***Helmholz resonator***; ***side-branch resonator***.

quartz halogen lamp High-intensity lamp or bulb of quartz glass containing an incandescent halogen gas such as iodine.

quartz iodine lamp Also ***tungsten halogen lamp***. See ***quartz halogen lamp***.

quench The cooling of a portion of the cylinder head gases during combustion in an IC engine, usually by minimizing the clearance between ***piston crown*** and ***cylinder head***. The high local surface:volume ratio effects rapid heat loss and suppresses detonation, though at the expense of higher ***hydrocarbon*** emission. See also ***squish***.

quench motor Engine exhibiting quench features in ***combustion chamber*** design.

quick-detachable coupling Quick-release air pressure hose connector for use with single line braking systems of ***trailers*** and ***semi-trailers***. See also ***contact coupling***; ***dummy coupling***; ***stand pipe***.

quick release valve (1) Valve that enables brake chambers of an ***air brake*** system to be rapidly emptied to effect the quick release of brakes. (2) Control unit which accelerates the release of air pressure from various portions of the brake system.

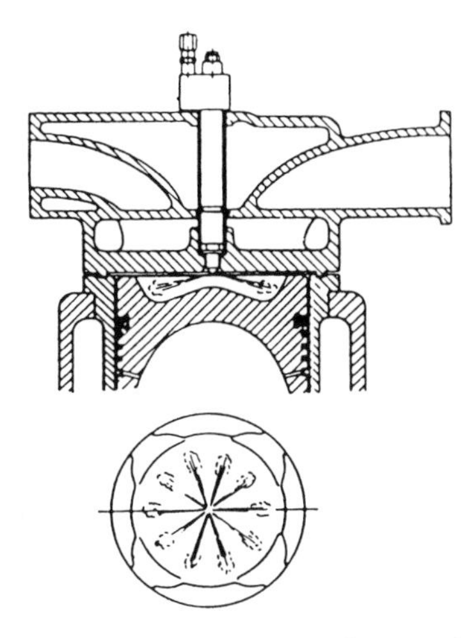

Figure Q.1 The quiescent combustion chamber of a larger diesel engine.

quiescent chamber Type of combustion chamber of a direct injection diesel engine, in which the mixing of the charge is predominantly non-turbulent or quiescent. See also ***quiescent combustion***. See Figure Q.1.

quiescent combustion Non-turbulent combustion, particularly in a lower-speed diesel engine, and normally employing a multi-nozzle injector.

quietening ramp Sector of engine valve train cam profile that reduces acceleration and so limits the generation of mechanical noise.

quill drive Mechanical drive system using concentric shafts mounted on separated bearings. The inner shaft is usually of sufficiently small diameter to provide torsional flexibility.

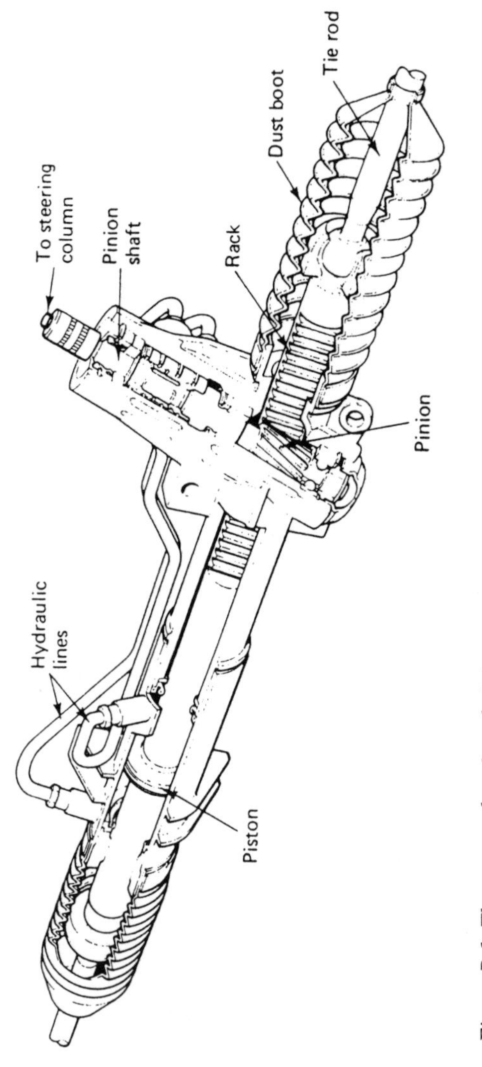

Figure R.1 The powered rack and pinion steering unit of a Volvo passenger car.

R

RAC horsepower Obsolete method of rating engines for taxation purposes using a formula based on cylinder bore and number of cylinders, instigated by the Royal Automobile Club. The rating was spuriously expressed as a horsepower, which it approximated when the formula was introduced. In the US the same method of rating was unjustly attributed to the SAE.

RAC rating See ***RAC horsepower***.

rack A straight length of toothed gearing.

rack and pinion Steering gear system in which a pinion on the end of the steering shaft engages with a horizontal rack, the ends of which are coupled to the ***steering arms*** by ***tie rods***. See Figure R.1.

radial (1) Disposed at right angles to an axis. (2) A ***radial ply tire***. (Informal)

radial engine Multi-cylinder engine in which the cylinders radiate from one single-throw crankshaft. For reasons of balance the number of cylinders is normally odd, three being the minimum to qualify for the definition. Widely used in aviation, rarely in automotive applications.

radial ply tire Tire in which the ***plies*** run radially, nominally at right angles to the ***bead***. Also ***radial tire***; ***a radial***. (Informal) See Figure R.2.

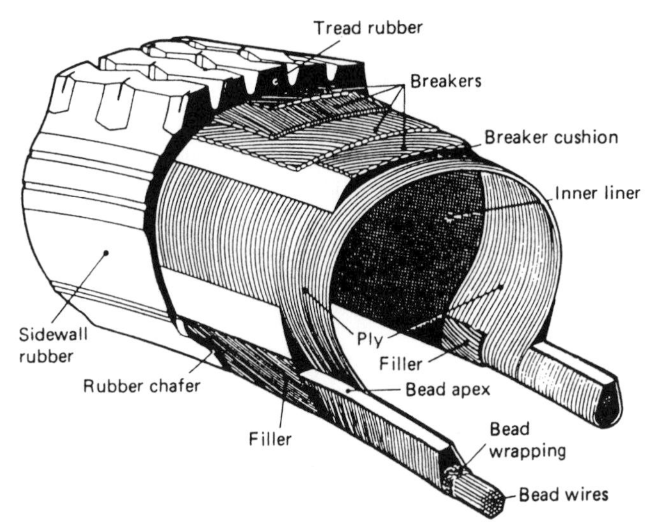

Figure R.2 Construction of a radial ply tire.

radial runout Condition of wheel and tire assembly when radius of tire tread from axle varies. An out-of-round condition. See also ***lift***.

radial truth See ***radial runout***.

radiator (1) Part of an engine cooling system through which excess combustion heat is lost to atmosphere by means of forced convection (not primarily by radiation as the name suggests) using a circulating liquid such as water or water/glycol to effect heat transfer. (2) Device for heating a vehicle interior as part of a vehicle heating system. See also ***oil cooler***; ***water pump***.

radiator blind (US: radiator shutter) Device that limits the flow of air through a radiator, as in conditions of extreme cold.

radiator cap Cap that closes header tank of radiator. Also ***pressure cap***.

radiator grille Protective and decorative guard to conceal radiator and protect it from impact by road debris while allowing free passage of cooling air. Occasionally ***grill***.

radius arm Suspension linkage providing fore and aft location of an axle. Also ***radius arm***; ***radius rod***; ***radius stay***; ***torque arm***; ***torque rod***. See also ***leading arm***; ***trailing arm***. See Figure W.2.

radius rod See ***radius arm***.

radius stay See ***radius arm***.

rag top (UK: soft top) Open car with stowable fabric roof and rear light. (Informal) Also ***convertible***.

rail A dragster. (Slang)

rainflow method A technique of fatigue analysis, analogous to rain flowing off a pagoda roof. See also ***trough method***.

rake angle See ***caster angle***.

ram air Air in orderly rapid motion relative to the vehicle and therefore possessing kinetic energy, particularly in context of air required to charge an air-breathing device such as a cooling system or engine induction system. Note that on the road the air is stationary (with no wind) and that it is the vehicle that moves.

ram air induction Pressure charging using the momentum of the induction air in an intake pipe of tuned length. See also ***harmonic induction engine***; ***tuned intake pressure charging***.

ram air supercharging See ***ram air induction***.

ram duct In an engine cylinder with two ***inlet valves*** of specific function, the part of the inlet tract that provides ram air, usually without intentional ***swirl***. See also ***swirl duct***.

ram pipe supercharging See ***ram air induction***.

ramp angle Angle to the ***ground plane*** of an abrupt incline or ramp. See Figure L.3.

range change Change-speed transmission for heavy vehicles in which the number of available speeds is doubled by the addition of an auxiliary gearbox (transmission) in series with the main gearbox, engagement of which provides a second range of gear ratios. Sometimes combined with a ***splitter*** to give a further set of ratios. See ***range-splitter***.

range-splitter Heavy-duty transmission gearbox incorporating both ***range change*** and ***splitter*** facilities to provide a larger number of close gear steps.

Rankine cycle Thermodynamic cycle of the steam engine and turbine involving evaporation of steam from water at boiler pressure, adiabatic expansion in cylinder or turbine, and condensation to the original temperature and pressure conditions. The practical steam engine deviates considerably from the ideal cycle in operation.

rape oil Vegetable oil sometimes used as a ***fuel***, and obtained by crushing the seeds of the rape plant. Also known as ***rapeseed oil*** and ***colza oil***.

rapid transit Any vehicle intended for rapid transportation of passengers, particularly in urban and suburban areas.

rated horsepower See ***rated power***.

rated power Power of an engine as stated in accordance with an acknowledged standard. See ***brake power***.

rated speed Engine speed at which rated power is obtained. Sometimes maximum engine speed or governed speed.

rave Stiffening and structural member that forms the edges of the load-carrying platform of a commercial vehicle or trailer. See Figure C.13.

rave hook Looped webbing hook, often of a folded wire rod eye form, for fastening the curtains of a ***curtain-sider*** vehicle. See Figure C.13.

RE valve See ***relay emergency valve***.

reach The significant length or extent of a component, for example the reach of a ***spark plug*** being the length required to locate the electrode correctly in the ***combustion chamber***.

reaction member Any suspension member that reacts a load or force, and provides positive location for another member. See also ***reactor***.

reactive suspension Tandem or multi-axle suspension in which a linkage between the axles or their suspension systems results in interactive response to transmission and braking inputs or bump displacement. See also ***balance beam***; ***compensating axle suspension***; ***non-reactive suspension***; ***walking beam***. See Figure T.1.

reactor The static reaction member of a ***torque converter***. See Figure T.3.

rear axle The ***back axle***.

rear end roll steer Steering effect imparted to a cornering vehicle by the ***roll deflection*** of the rear suspension.

rear engine Engine mounted behind the rear wheels of a vehicle.

rear fender (UK: rear wing, tonneau panel) Body panel of passenger car between rear door and end of vehicle, and enclosing rear wheel. See Figure B.4.

rear lamp Red lamp at extreme sides of back of vehicle to indicate presence and width. See also ***tail lamp***.

rear quarter panel The panel adjacent to the rear door and ***D-post***. See Figure B.4.

rear spoiler Transverse aerofoil or projecting surface mounted at the rear of a vehicle to reduce ***lift*** and ***drag***.

rear underrun bumper Structure at rear of goods vehicle to prevent intrusion of other vehicles in rear impact. See Figure L.3.

rear-view mirror Internal mirror to provide driver with rearward view through the backlight.

rear vision area Portion of the backlight used by the inside ***rear-view mirror*** for vision to the rear.

rear-wheel drive Transmission system in which the engine power is delivered to the rear wheels of a vehicle. See also ***Hotchkiss drive***.

rebore Regrinding of worn or damaged cylinder bore of an engine (or pump) prior to the fitting of oversize piston, usually as part of a reconditioning process.

rebound Extension of suspension travel beyond static condition.

rebound clip Metal clip to prevent separation of leaves of a ***multi-leaf spring*** on ***rebound***.

rebound stop Device for increasing the spring rate towards the end of rebound travel.

recapped tire A retreaded tire. (Informal)

reciprocating engine An engine in which a piston is constrained to move to-and-fro in a cylinder by a ***crankshaft*** and ***connecting rod***, or other mechanical arrangement.

recirculating ball steering Type of ***worm and nut*** steering in which recirculating steel balls occupy the space between the nut and worm wheel, thus reducing friction.

reciprocating seal Seal between one fixed component and a reciprocating shaft, as for example a ***valve stem*** seal.

recoil starter Engine hand starter in which the energy for starting is stored in a coiled spring, or which uses a spring to rewind a hand pulley-cord.

reconditioned engine Used engine renovated for continued use by rebore, regrinding of crankshaft journals and replacement of items subject to wear or or other forms of degradation. Also ***recon***. (Slang) See also ***short engine***.

recovery vehicle (US: wrecker) Salvage vehicle equipped to retrieve a damaged or stranded vehicle. Also ***breakdown truck***.

recreational vehicle Vehicle intended primarily for off-highway leisure use. Also ***RV***.

rectifier Electrical device for changing alternating to direct current.

recut tire Used tire in which the existing tread has been cut deeper to extend the tire's useful, or legal, life. See also ***remold tire***.

redline Segmental mark on an instrument, such as a speedometer or tachograph, indicating a maximum that should not normally be exceeded.

reed valve One-way valve in which sprung plates or fingers deflect under pressure to admit a working fluid. Sometimes used in two-stroke motorcycle engines and in air compressors.

reefer A refrigerated, insulated box van. (Informal) See Figure R.3.

Figure R.3 A forward-control refrigerated vehicle, or "reefer."

refining Process or processes whereby ***crude oil*** is converted to usable products.

reflector See ***reflex reflector***.

reflector panel An external panel that reflects incident light.

reflex reflector Molded optical mirror to indicate presence of a vehicle by reflecting incident light close to the direction of source.

reformulated gasoline Gasoline formulated to minimise exhaust emissions, having for example low vapor pressure and low ***aromatic*** and sulfur content, with the addition of oxygenates to reduce ***emissions*** of carbon monoxide and unburned hydrocarbons.

refrigerated vehicle An insulated box van equipped with a refrigeration unit for conveyance of foodstuffs and other cargoes requiring conveyance at low temperature. See also ***chilled distribution***. A ***reefer***. (Informal) See Figure R.3.

refuse vehicle (US: garbage truck) Vehicle for collecting and compacting waste. ***Dust cart***. (UK informal)

regenerative braking Braking in which the loss of kinetic energy from braking is stored and subsequently fed back to provide tractive effort. Currently practical only in electric vehicles where braking can be aided by switching the motor to generator mode, and energy stored in the battery.

regenerator A form of heat exchanger.

regrooved tire Used tire on which a new ***tread*** has been cut, either into the original tread or on a retread, but to a greater depth than that of the original molded groove.

regulator (1) Electrical, electronic or semi-conductor assisted device for stabilizing or controlling the current and voltage in a vehicle electrical system. See also ***current regulator***; ***cut-out***; ***electronic regulator***; ***voltage regulator***. (2) The lever mechanism by which a vehicle window is raised or lowered.

regulator handle The manually operated handle by which a vehicle window is raised or lowered. See also ***regulator***.

reinforcing plies Plies laid circumferentially beneath the tread of a tire. Also ***undertread***. See Figure R.2.

relative wind The velocity and direction of ambient airflow experienced by a moving vehicle, consisting of the components of vehicle velocity and wind velocity and direction.

relay (1) Electrical switch magnetically operated by flow of current through a coil, used prior to electronic systems for voltage and current controllers. (2) Any unit, mechanical, electrical or otherwise, that remotely actuates or switches in response to a signal or stimulus.

relay emergency valve Valve that applies trailer brakes in the event of a breakaway. Also ***RE valve***.

relay lamp (UK: additional direction indicator lamp) Additional flashing traffic indicator, particularly at side of long vehicle.

relay rod A ***track rod*** or part of the track rod directly coupled to the ***drop arm*** or ***pitman arm***. Also ***intermediate rod***.

relay valve Valve situated near local reservoir and brake chamber of an ***air brake*** system to decrease the response time of the brake, particularly on long vehicles.

release bearing See ***clutch release bearing***; ***throwout bearing***; ***throwout sleeve***. See Figures C.4 and C.5.

release lever See ***clutch release lever***. See Figures C.4 and C.5.

release lever plate Clutch element that communicates the axial motion of the ***clutch release bearing*** to the ***clutch release levers***. Also ***clutch release lever***. See Figure C.4.

reluctor Electronic spark triggering device in an ***electronic ignition*** system.

remold tire (UK: remould) Used tire onto which a new tread has been molded, the worn tread having been removed. See also ***recut tire***; ***retreaded tire***.

removal van (US: moving van) Large-capacity van for removal of household effects. See also ***pantechnicon***.

re-refining The reprocessing of used ***oils*** to produce stock of near to the original properties.

Research Octane Number A guide to anti-knock performance of a fuel under mild driving conditions. It is derived from one of the standard comparative tests using the Cooperative Fuel Research (CFR) engine. Also ***RON***. See also ***Motor Octane Number***.

resin Solid or semi-solid natural or synthetic organic material used in manufacture of adhesives, some paints, and molded components. See also ***acrylic***; ***polyester***.

resonator Acoustic chamber, generally of fixed volume, with a specific resonant frequency. Applications include induction and exhaust systems. See also ***side branch resonator***.

restraint system (1) Means of holding in place against the action of loads, particularly loads resulting from acceleration or deceleration. (2) A safety restraint system such as a ***seat belt*** or ***air bag***.

retaining ring (1) Any annular fitting for holding a movable item securely in place. (2) Clamping device for holding a headlamp in position.

retard stop Mechanical contact to restrict degree of ignition ***retard*** of a ***centrifugal advance*** distributor. See also ***advance stop***.

retarded ignition Timing of the ignition spark to occur later than the optimum timing for fuel of specified ***octane rating***. Ignition may be retarded to enable an engine to operate on fuel of lower than specified quality. See also ***advanced ignition***.

retarder Any device that supplements a vehicle's ***service brakes*** by dissipating energy of motion, as by electrical or hydraulic means. Also ***brake retarder***. See also ***electric retarder***; ***engine brake***; ***exhaust brake***.

retreaded tire Used tire to which a new tread has been affixed by vulcanization to extend its useful life. Also ***retread*** (Informal); ***remould*** (UK); ***remold*** (US). See also ***regrooved tire***.

rev counter A ***tachometer***. (Informal)

reverberation chamber Acoustic test chamber, the walls of which are largely reflective of sound, as for the execution of noise fatigue tests. Also ***reverberation room***.

reverberation time Time, usually in seconds, that a stipulated sound source takes to decay by a specified amount (often 60 dB) after cut-off.

reverse Ackermann Steering geometry which applies the opposite effect to ***Ackermann steering***, with the outer wheel turning at a greater angle than the inner.

reverse flow See ***cross scavenging***.

reverse gear Gear that changes the sense of rotation of the output shaft with respect to the engine, thus permitting rearward motion of the vehicle.

reverse rod Rod that connects control lever to the reverse gear ***selector mechanism*** in an ***automatic transmission***.

reversed clutch Clutch assembly with input and output shafts on the same side, as for a transverse engined car.

reversed Elliot Arrangement of steered ***stub axle*** whereby the axle beam terminates in an eye end which holds the ***kingpin***. The stub-axle kingpin axis divides as a yoke or fork-end which straddles the axle beam end. See also ***Elliot axle***; ***Lemoine***. See Figure R.4.

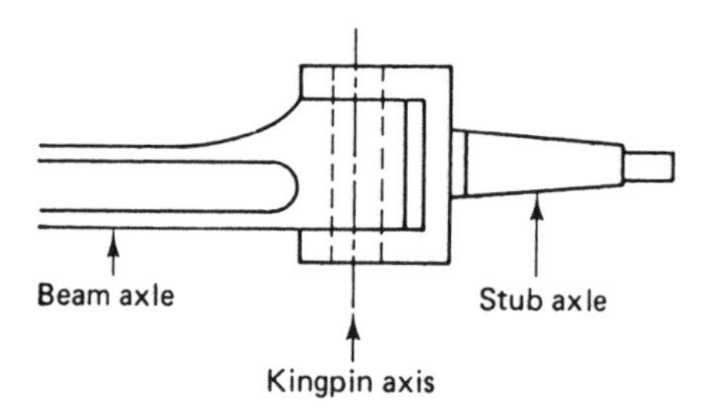

Figure R.4 Reversed Elliot steering arrangement of a beam axle.

reversed offset wheel Wheel in which the ***nave face*** is inboard of the tire center plane. See also ***offset wheel***.

reversible kingpin A semi-trailer kingpin which is reversible or interchangeable to suit different types of ***fifth-wheel*** coupling.

reversing bleeper (US: back-up alarm) Acoustic warning that automatically sounds when a vehicle, particularly a commercial vehicle, engages reverse gear.

reversing lamp (US: back-up lamp) Lamp to illuminate to the rear of a vehicle.

revolution counter See ***tachometer***. Also ***rev counter***. (Informal)

rheology The study of the flow of ***greases*** and other semi-solid materials.

rhombic drive Mechanism that converts reciprocating to rotary motion when two pistons are required to operate out of phase on a common axis. A feature of some ***Stirling engines***. See Figure R.5.

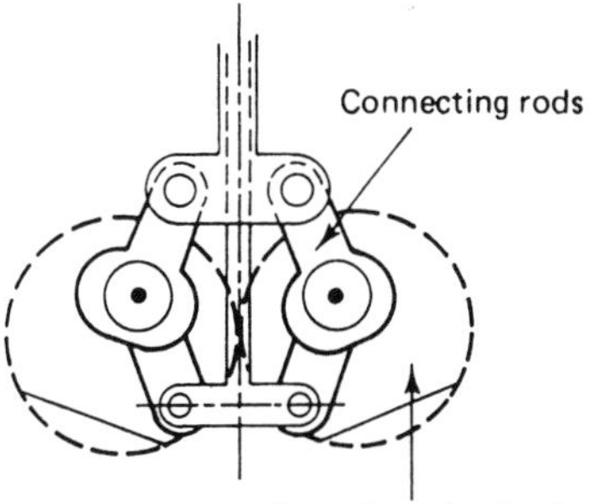

Figure R.5 Rhombic drive crank and connecting rod arrangement of a Stirling engine.

rib Raised circumferential feature of a tire tread pattern. See also ***void***.

Ricardo engine A standard variable compression engine used for test or research purposes.

Ricardo head Type of high ***squish***, high turbulence combustion chamber for ***side-valve (L head)*** engines. See Figure S.5.

rich mixture Fuel/air mixture in which the proportion of fuel exceeds that necessary for theoretically complete or ***stoichiometric*** combustion.

ride Qualitative and quantitative nature of the response of a vehicle to road perturbations, particularly in relation to the performance of the suspension system and its influence on the degree of comfort experienced by the occupants.

ride clearance Maximum displacement in compression of a suspension unit from the normal load position. Also ***ride height***.

ride height Attitude of a vehicle on its suspension, whether heavily or lightly loaded. Often used qualitatively rather than quantitatively.

ridging (US: nibbling) Effect of lateral force on a ***tire*** resulting from glancing contact with a rut or shallow ridge.

right-hand drive Controlled from the right-hand side (starboard) of the vehicle.

rigid See ***rigid truck***; ***rigid vehicle***. (UK informal)

rigid axle A ***beam axle***.

rigid truck A non-articulated vehicle, particularly a commercial vehicle carrying the payload on its own axles. See Figure R.6.

rigid vehicle Vehicle which serves its function as a unit and not as a part of an articulated vehicle or train. Thus the tractor unit of an articulated vehicle is not, by many definitions, a rigid vehicle. See Figure R.6.

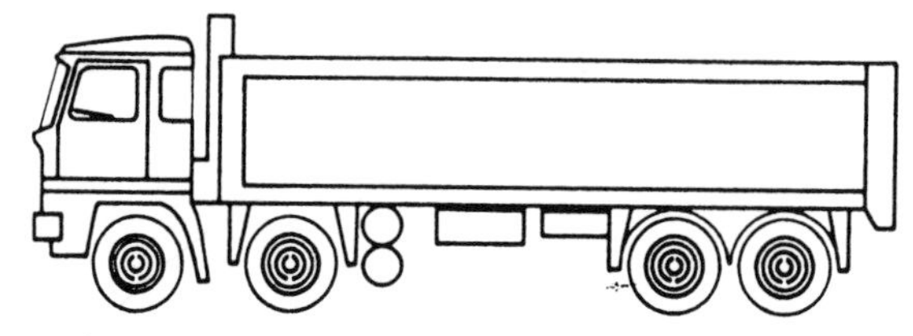

Figure R.6 A rigid eight-wheeler tipper truck.

rim (1) Part of a wheel assembly on which the ***tire*** is mounted and supported. (2) The outer edge or a circular component.

rim flange The retaining flange of a wheel rim, which supports and retains the ***tire bead***. See also ***bead seat***. See Figure W.5.

rim pull Tractive effort of a vehicle as measured at the wheel/road interface, or as calculated from torque and transmission ratio data.

rim wheel Open structure fabricated ***tractor*** wheel with broad annular rims usually fitted with removable lugs or "spades."

rim width The width across a wheel, measured between the rims. See Figure W.5.

ring expander Annular spring within ***piston ring*** to enhance sealing action of lower rings and in worn cylinder bores.

ring gap Gap in the annular shape of a ***piston ring***, enabling the ring to be opened out for installing, and to accommodate thermal expansion.

ring gear (1) Spur gear teeth set on periphery of ***flywheel*** with which the ***starter motor*** pinion engages. See Figures C.4. and E.1. (2) Any large-diameter annular gear wheel, whether internally or externally toothed. See also ***annulus***.

ring groove Annular slot or recess in piston into which a ***piston ring*** fits.

rising rate suspension Suspension in which the spring rate increases with compression.

road feel Feedback of front suspension and steering input to ***steering wheel***, and to driver.

road octane number Usually the octane number of a ***primary reference fuel (PRF)*** that just gives trace knock in a vehicle on the road or on a ***chassis dynamometer*** under specified conditions.

road plane Reference plane on which a vehicle stands. This may be inclined to represent road camber or surface curvature, subject to local definition. See also ***ground plane***.

road tanker Commercial vehicle equipped to convey liquids or bulk powder. See Figure A.4.

road train Heavy goods vehicle towing one or more trailers. See also ***locomotive***.

roadholding Qualitative and non-technical term for handling, particularly freedom from oversteer.

roadster Passenger car of sporting appearance. A ***convertible***. (US informal)

roadway flusher Vehicle equipped for washing roadways. A watering cart.

rocker See ***rocker arm***.

rocker arm (1) Centrally pivoted beam which transmits a linear displacement at one end to a linear displacement at the other. (2) In particular, the part of the ***valve train*** of an engine that transmits the ***lift*** of a ***cam*** to operate a ***poppet valve***. See also ***Roesch rocker***. See Figures C.14 and E.1.

rocker arm cover (UK: rocker box) Cover for the ***rockers*** and associated valve gear of an engine, attached to the ***cylinder head*** of an engine.

rocker shaft (1) Shaft about which ***rocker arms*** pivot in a ***valve train***. See Figure C.14. (2) A steering ***sector shaft***.

rod (1) A connecting rod. (Informal) (2) A car tuned for performance or otherwise personalized. (Slang) (3) To scour or clean pipework of a cooling or hydraulic system.

rod bearing See ***connecting rod bearing***.

Roesch rocker Pressed ***rocker arm*** of concave section, pivoting on a semi-spherical washer with nut to facilitate adjustment rather than on a rocker arm shaft. Also ***ball-type rocker***. See Figures E.1 and W.3.

roll Angular displacement of the sprung mass of a vehicle about its longitudinal axis.

roll axis The line joining the ***roll centers*** of the front and rear suspension of a vehicle. The axis about which a vehicle rolls in response to a roll perturbation.

roll bar (1) Single element transverse frame to protect occupants of an open car in event of a ***roll-over***. Also ***roll-over bar***. (2) An anti-roll bar in a suspension system. (Informal) See Figure A.3.

roll cage Safety structure built onto an open or ***convertible*** car, agricultural tractor or off-highway vehicle to protect occupants in event of a ***roll-over***.

roll camber Change of wheel ***camber*** resulting from suspension roll.

roll center Instantaneous geometric center about which a vehicle body rolls, as determined by the suspension geometry. Normally, both axles will have their own roll centers, the line joining these determining the ***roll axis***. See Figure S.11.

roll-over The complete capsizing of a vehicle.

roll oversteer Roll steer which increases ***oversteer*** or reduces ***understeer***.

roll rate (1) Suspension roll stiffness. (2) Vehicle roll angular velocity, or roll frequency.

roll steer Change of ***steer angle*** due to suspension roll.

roll stiffness Resistance of a vehicle to rolling, expressed mathematically as a moment per unit angular displacement in roll.

roll understeer Roll steer which increases ***understeer*** or reduces ***oversteer***.

roll velocity Angular velocity about the roll axis.

roller chain Chain in which the fastening pins carry rollers, permitting high speeds of operation, with reduced wear and noise. Used for power transmission as in camshaft drives, motorcycle and bicycle transmissions.

roller clutch ***Freewheel*** or ***one-way clutch*** using hardened balls in tapered detents.

roller lifter Cam follower, lifter or tappet incorporating a small wheel which is in rolling contact with the cam. Rare, except in some larger diesel engines.

roller mat conveyor System of rollers built into the floor of a box-van or trailer to facilitate the movement of heavy or palletised loads.

rolling element bearing A bearing in which the relative motion of the two surfaces is made possible by the rolling of spherical, cylindrical or conical elements, such as a ball bearing, roller bearing or needle-roller bearing. See also ***plain bearing***.

rolling radius Radius of loaded tire from axis to center of ***tire footprint***.

rolling resistance Retarding effect on a vehicle of its motion on a flat road surface, consisting primarily of rolling and mechanical friction, but excluding air resistance. See also ***driving resistance***; ***tire resistance***.

rolling road Test apparatus consisting of horizontal powered revolving cylinders on which the road wheels of a vehicle ride to facilitate laboratory observation of the performance of a vehicle, its engine and transmission, under simulated road conditions. A ***dynamometer***; ***brake tester***. (UK informal)

RON See ***Research Octane Number***.

roof bow See ***roofstick***.

roof lining (UK: headlining) Fabric or soft material roof ceiling of a vehicle passenger compartment or cab.

roof panel The (usually) metal pressing that makes up the roof of a car or cab. See Figure B.4.

roof rack Framework, usually detachable, for carrying items on a vehicle roof.

roofstick (US: roof bow) Transverse frame or stiffener supporting a vehicle roof, but particularly the curved roof of a box body van or bus.

Roots blower Mechanical induction pressure charger in which air pressure is provided by the contiguous rotation of two or three lobed rotors or pistons. Not Rootes.

rose joint Proprietary eye-end or ball and socket joint incorporating a spherical bearing.

rotary engine (1) Engine that converts the energy of gas expansion to mechanical power by direct rotary as opposed to indirect reciprocating action. See also ***Wankel engine***. See Figure W.1. (2) Obsolete form of ***radial engine***, used mainly for aircraft propulsion, in which the crankshaft remained stationary while cylinders and airscrew rotated about the crankshaft axis.

rotary valve Valve system of an engine or pump in which the valve ports form part of a rotating assembly, usually of tubular or disc configuration.

rotating beacon Any light system that displays a rotating beam of light in a nominally horizontal plane. Also ***rotating warning lamp***; ***warning light system***.

rotating shaft seal Seal, usually molded from an ***elastomer***, to prevent loss of fluid, as for example from a ***gearcase***.

rotating warning lamp See ***rotating beacon***.

Rotoflex joint Proprietary ***universal joint*** in the form of a hexagonal rubber ring.

rotor (1) Component or assembly of components that turns about its own axis, as of a turbine or electric motor. (2) The rotating part of a rotary engine, such as a ***Wankel engine***, that converts gas expansion into rotary motion. (3) The rotating part of an axial compressor or gas turbine that carries (and includes) the blades. (4) The contact arm of the ***distributor*** in a spark ignition engine. Also ***rotor arm***.

rotor arm Rotating ignition ***distributor*** component that establishes high tension electrical contact between source and the individual spark plug HT leads.

roughness Vibrations up to 100 Hz, generated by a tire on a smooth road surface, but producing the sensation of traveling on a coarse surface.

rubber Strictly informal but widely used term for elastometic materials, and particularly for natural rather than synthetic elastomers, which are nevertheless widely known as synthetic rubbers.

rubbing strip Longitudinal rubber or plastics molding attached at ***waistline*** of vehicle to prevent damage through contact or minor impact.

Rudge nuts Quick release nuts for releasing wheels, particularly on racing cars. (Obsolete) See also ***knock-on wheel***.

rumble (1) Low-pitch thudding, differing from ***knock*** and accompanied by engine roughness, usually caused by high pressure rise rate associated with very early ignition systems or ***surface ignition***. (2) Low-frequency tire-generated noise, characteristic of certain surfaces.

rumble seat (UK: dickie seat) An occasional open tail seat.

run-flat tire Pneumatic tire capable of limited operation without air inflation.

run-in (US: break-in) To run new or reconditioned machinery under light load for a preliminary period. See also ***running-in***.

run-on Continued firing of a spark ignition engine after the ignition system has been switched off. Also ***after-running***. See also ***dieseling***.

run-on tire Pneumatic tire designed for use with special rims to prevent bead unseating in the event of a ***puncture*** or ***blow-out***, thus enabling a moving vehicle to be brought safely to rest.

runaway knock Engine knock that becomes progressively worse under steady speed and load conditions.

runaway surface ignition Improper ignition within engine cylinder emanating from hot engine components (not from carbon deposits) and becoming more pronounced as engine temperature increases.

running board Lengthwise horizontal step at ***sill*** level to facilitate access to a vehicle.

running gear (1) The driving, steering and suspension mechanism of a vehicle. This term often implies the unsuspended undercarriage components such as ***wheels*** and ***axles***, ***final drives*** and ***steering linkages***. (2) The undercarriage of a vehicle.

running-in (US: breaking-in, wearing-in) Initial operation of new or rebuilt machinery under light load and reduced speed to ensure even bedding of bearings and reciprocating items.

runout Eccentricity, and in some cases allowable eccentricity, of a shaft, bearing or other rotating component.

rust inhibitor (1) Temporary protective coating for metal. See also ***rust preventive***. (2) Chemical ***additive***, for example to a lubricating oil, that inhibits the rusting of ferrous components, such as gears or bearings. Special inhibiting oils are often used prior to putting machinery temporarily out of commission, as for winter storage.

rust preventive Liquid compound, sometimes in the form of a varnish, used for coating surface of steel to prevent rusting in storage. See also ***rust inhibitor***.

RV See ***recreational vehicle***.

Rzeppa joint ***Constant velocity joint*** in which driving and driven members bear spherically upon caged balls.

S

S-cam brake Drum brake with S-shaped cam to actuate the ***shoes*** and compensate for ***lining*** wear. See Figure C.1.

sac Cavity immediately before the ***discharge orifice*** of a diesel ***fuel injector***. See Figures I.3 and P.1.

sac volume Volume of the ***sac*** of a diesel ***fuel injector***.

saddle (1) The molded seat for the rider of a bicycle or motorcycle. (2) The flat or concave support and attachment point of an axle on which a ***leaf spring*** is mounted, often with an intermediate ***cushion***. (3) Any concave or flat component or surface that provides positive location for the attachment of an item. See also ***seat***.

SAE rating Standard of engine horsepower measurement using a bare engine, thus excluding losses to ***generator***, pumps and other ancillaries. See also ***DIN rating***.

SAE viscosity Viscosity of motor ***oil*** according to system developed by the SAE.

safety belt See ***seat belt***.

safety bumper Bumper designed to protect vehicle on impact or reduce injury on impact with other road users. See also ***Federal bumper***.

safety cell Occupant accommodation space as a structural unit which resists ***impact*** collapse or intrusion.

safety glass Glass that is resistant to impact and that shatters in such a way as to minimise injury on breakage or when impacted by vehicle occupant. See also ***laminated glass***; ***toughened glass***; ***Triplex***.

safety glazing Transparent material of vehicle windows that does not shatter into sharp fragments, and otherwise minimises injury. Term more often used of synthetic materials rather than glass.

saloon (1) Passenger compartment of a bus or coach. (2) Saloon car (UK) is equivalent to ***sedan*** car (US).

salt spray test Test for a vehicle's resistance to corrosion.

sandshoe Trailer ***landing leg*** foot for use on soft or uneven terrain.

scatter shield Protective safety cover over any item liable to explosive failure, such as the ***clutch*** assembly of a racing car.

scavenge filter Filter in the return line of a ***dry-sump*** lubrication system to remove metallic debris and other contaminants from the oil before its return to the oil tank.

scavenge plunger See ***scavenge pump***.

scavenge pump (1) Oil pump that returns oil to the main oil reservoir in a ***dry sump*** lubrication system. Also ***scavenge plunger***. (2) Any pump that serves to remove unwanted fluids or suspensions from a system.

scavenger ***Halogens*** present in a lead ***anti-knock*** compound which prevent the build-up of lead oxides and sulfates in an engine's ***combustion chamber***, by rendering lead compounds formed during combustion volatile, so that they pass through the exhaust.

scavenging Removal of exhaust gases from an engine cylinder, particularly by an induced flow of gas.

scavenging efficiency The ratio of the mass of working gas retained in an engine ***cylinder*** at completion of the exhaust cycle to the total charge mass supplied. See also ***volumetric efficiency***.

Schnuerrle system System of scavenging in ***two-stroke engines***, also called ***loop scavenging*** or reversed loop scavenging, in which the scavenging flow enters and leaves the cylinder from the same side, having performed a loop from ***piston crown*** to ***cylinder dome***.

Schrader valve Type of pneumatic non-return valve used especially for ***tire*** inflation. Depression of a central spigot opens the valve to facilitate inflation and the measurement of tire pressure.

scoop A cowled aperture to gather and direct ambient air, generally for cooling. See also ***NACA duct***.

Scotsman's sixth Practice of coasting downhill with gears disengaged, particularly of a commercial vehicle. (Mainly UK usage) Also ***Mexican overdrive***.

scouring Form of ***tire*** wear caused by inaccurate tracking.

scout car A small military vehicle, mainly for reconnaissance and personal transportation.

scraper Mobile earthmoving machine equipped with a rigid transverse horizontal blade generally set well forward of a trailer axle, for cutting, leveling and removing soil, which may be loaded into an integral hopper or bowl.

scraper ring Piston ring that removes excess oil from a cylinder bore and returns it to the lubricating circuit. Also ***oil control ring***. See Figure P.2.

screen Transparent panel for dividing compartments of a vehicle or deflecting airflow. Also light, as in ***backlight***. A ***windscreen*** or ***windshield***.

screen wiper Windscreen or windshield wiper. (UK informal)

screw and nut steering See ***worm and nut steering***.

scuffing Abrasive or adhesive damage to surfaces in relative motion resulting in scraping or scratching, for example of cylinder bore journal bearings, piston rings, cams and tappets. See ***pitting***.

scuttle The lower, forward part of a driver's cab or passenger compartment that accommodates the legs of front-seated occupants and forms a ***bulkhead*** or ***firewall*** with the engine compartment. (UK informal)

seal swell Increase in volume of certain rubbers due to the action, for example, of ***lubricating oil*** or ***hydraulic fluid***.

sealed beam Lamp, such as a ***headlamp*** or ***fog lamp***, of which all components such as lens, reflector and light source are an inseparable unit.

seat (1) Support and attachment point, particularly where load carrying, as for example a spring seat. (2) The seat of a driver or passenger. See also ***bucket seat***; ***squab***. (3) The seating of a mechanical item, such as a ***valve*** seat.

seat belt Occupant restraint system consisting of webbing and latching hardware, and firmly anchored to the vehicle structure.

seat belt grabber Device that prevents the unwinding or paying-out of a ***seat belt*** as it tightens on the reel or drum on impact. See also ***anchorage***; ***secondary safety***; ***sisters***.

seat belt pre-tensioner Device that actively pre-tensions a ***seat belt*** on impact by the action of a pre-tensioned spring, small explosive charge, or similar means.

seat pan Transverse horizontal structural panel on which seats, usually only the rear seats in a passenger car, are placed.

secondary In engine balancing, an event occurring at twice engine speed. See also ***secondary couple***; ***secondary force***.

secondary battery An electrical battery that can be recharged, as can a ***lead-acid battery***.

secondary brake Heavy vehicle ***spring brake*** system which, when air pressure is released, allows the spring brake to exert force, and particularly to provide a positive brake force for parking. Air pressure failure will also bring this brake into operation.

secondary couple The out-of-balance disturbance in an engine that causes rocking about a horizontal or vertical axis (in a vertical or V configuration engine) at right angles to the ***crankshaft***, and at twice engine speed.

secondary force The out-of-balance force (in any plane of the crankshaft axis) resulting from finite length of the ***connecting rods*** in a real system, and acting at twice engine speed.

secondary safety See ***passive safety***.

secondary shoe The ***trailing shoe*** in a ***drum brake***. See Figure C.1.

secondary venturi Small venturi mounted coaxially within the main venturi of a ***carburetor*** to provide higher air velocity. Also ***auxiliary venturi***.

sector shaft Output shaft from a ***steering box***, to which the ***drop arm*** or ***pitman arm*** is attached. Also ***rocker shaft***.

sedan Conventional enclosed passenger car. (Mainly US usage)

sedan delivery body (UK: car-derived van) Van or delivery vehicle based on a passenger car body and running gear.

sediment Solid, usually ***particulate***, material that settles out of a liquid in which it is carried.

sedimenter Device for isolating and/or removing sediment from a fuel or other liquid. See also ***agglomerator***; ***filter***; ***separator***.

seize Sudden adhesive or frictional locking of parts normally in lubricated sliding contact, due to surface welding or clamping, as of a piston in its bore. Also ***seizure***; ***seize-up***.

select To choose, and by implication to engage, as of a gear in a ***change-speed gearbox***. See also ***pre-selector***.

selector fork Forked member for moving a sliding pinion into and out of engagement in a ***change-speed gearbox*** or ***transmission***. See Figures G.1. and S.1.

selector lever See ***selector fork***.

selector mechanism Mechanism for selecting a gear in a change-speed transmission, comprising ***gear lever***, ***selector fork*** and associated shafts and bearings. See Figure S.1.

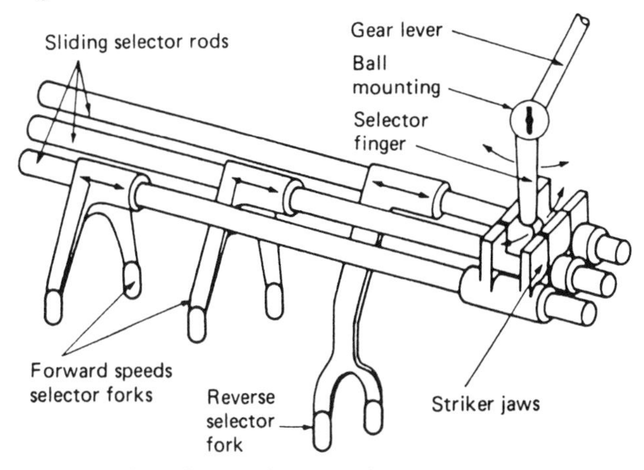

Figure S.1 Multi-rail gear selector mechanism.

selector rod Shaft on which a ***selector fork*** travels in a ***change-speed gearbox***.

self-adjusting tappet Tappet devised to automatically compensate for deviations in valve operating clearance, for example a ***hydraulic tappet***.

self-aligning torque Torque acting on the ***tire*** of a ***steered wheel*** that tends to reduce the slip angle and contributes to the feel experienced by the driver at the steering wheel. See also ***caster action***.

self-ignition See ***auto-ignition***.

self-indexing starter Four-pole four-brush starter of cranking motor with ***plate clutch*** mechanism.

self-leveling air suspension Air suspension in which the air pressure of each suspension unit can be varied so that the chassis remains level.

self-parking wiper Screen wiper that automatically aligns itself in its parked position when switched off.

self-steering axle Rear axle of the ***bogie*** undercarriage of a truck or trailer which pivots or casters as the vehicle changes direction. Also ***self-tracking axle***.

self-tracking axle See ***self-steering axle***.

semaphore indicator Vehicle direction indicator consisting of an arm on the side of the vehicle that can be manually or automatically raised and lowered. See ***trafficator***.

semi-anechoic chamber See ***hemi-anechoic chamber***.

semi-automatic headlamp beam switch Device which provides automatic or manual control of headlamp beam switching at the option of the driver. See ***headlamp beam switch***.

semi-automatic transmission Transmission system that requires some manual control, such as gear selection, but that engages the selected gear automatically. A ***pre-selector*** or ***Wilson gearbox***.

semi-centrifugal clutch Single-plate dry ***clutch*** in which the inter-plate pressure is augmented by levers with bob-weights to prevent slippage at higher speeds and torques. (Obsolete)

semi-continuous braking Braking of a combination of vehicles, such as a tractor unit and ***full trailer***, or an ***articulated vehicle***, where the operation of the brakes of the combination is supplied by two different sources, including the effort applied by the driver to the brake pedal. See also ***continuous braking***.

semi-diesel Compression ignition engine working on other than the ideal diesel cycle, as for example with hot-plug ignition.

semi-elliptical spring Leaf spring operating as a pin-ended beam reacting loads acting at its mid-length in bending, and so called originally because of its unloaded shape, though modern types approximate a shallow arc or catenary. May be of single or multi-leaf configuration. The traditional means of suspending a beam axle. Also ***half-elliptic***; ***semi-elliptic leaf spring***. See also ***multi-leaf spring***; ***parabolic spring***; ***single leaf spring***. See Figure L.2.

semi-floating axle Live axle assembly in which the weight of the vehicle is transferred from the axle casing or housing to the axle shaft, usually by a bearing within the housing. In a live semi-floating axle the axle shaft will carry rotational as well as bending loads. Also ***half-floating axle***. See also ***fully floating axle***; ***non-floating axle***.

semi-forward control Cab arrangement as of a commercial or public service vehicle, where the driver sits alongside the rear end of the engine. See also ***forward control*** and ***normal control***.

semi-tracked vehicle Vehicle, particularly a military vehicle, with traction provided by track-laying apparatus, but steered by conventional wheels. A ***half-track***.

semi-tractor Commercial vehicle tractor unit for hauling a ***semi-trailer***. (Mainly US usage)

semi-trailer Trailer of which the forward end when in transit is supported by and coupled to a tractor unit, normally by means of a ***fifth-wheel***, the combination of ***tractor*** and ***semi-trailer*** making an articulated vehicle. See Figure A.4.

semi-trailing arm Trailing ***suspension*** linkage, usually employed in independent rear suspensions, in which the ***pivot axis*** is inclined backwards in the horizontal plane to impart an increase in negative camber with increase of load. Also ***semi-trailing link***. See Figure S.2.

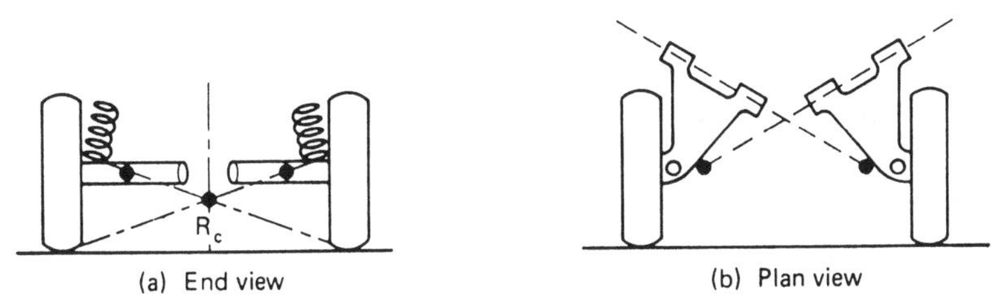

Figure S.2 Semi-trailing arm rear suspension showing position of roll center.

sender unit Transducer, usually electronic or electrical, that provides a signal related to a physical condition, such as temperature, pressure or rotational speed, to operate an instrument or warning light.

sensitivity (1) Of a control system, the degree of response to an input, as for example sensitive steering or a sensitive throttle. (2) Of a gasoline fuel, the difference between the ***research octane*** and ***motor octane*** numbers.

sensor Device in a control system that provides information from an input, as for example of rotational speed or line pressure.

separation Change of aerodynamic flow from laminar to turbulent.

separator Device for removing suspended particles from a fluid, or water from a ***fuel*** or ***oil***. See also ***filter***; ***centrifuge***; ***cyclone***.

service brake Primary brake used for retarding and stopping a vehicle.

servo A ***servomechanism***. (Informal)

servo-assisted Assistance, particularly of human effort, by mechanical power, as in servo-assisted brake, direct manual operation in some instances being possible without the aid of the ***servomechanism***.

servomechanism System or device that supplements physical effort in operating a control, either by direct mechanical amplification of feedback or by external power assistance. See Figure S.3.

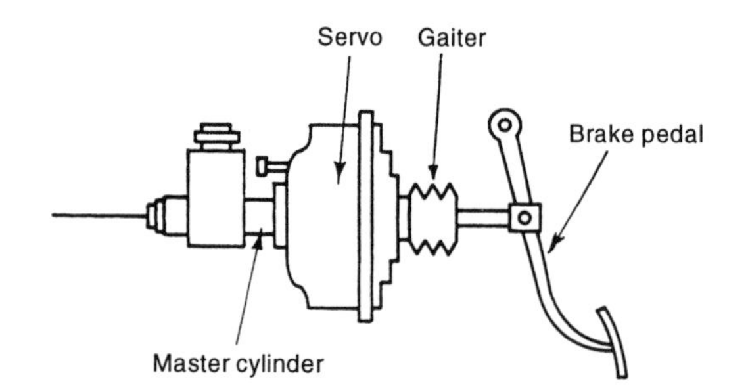

Figure S.3 A direct pedal-operated brake pneumatic servo.

set-back axle Vehicle running gear configuration in which the front axle is set substantially back from the front of the vehicle. See Figure R.3.

SFC See ***specific fuel consumption***.

shackle Pivoting link between a leaf spring and its mounting, comprising two coupled parallel rods. See Figure L.2.

shaft guard Safety guard to prevent injury from rotating shafts and couplings on machinery under test.

shaft whirl Resonant mode of a rotating shaft, such as an ***axle*** or ***crankshaft***, when deflection from its true axis is abnormally high. See also ***whirling speed***.

shaker Vibration and fatigue testing machine, usually of moving-coil electromagnetic type. (Informal)

shell A bare enclosed structure, but especially one made mainly of panels. A monocoque structure. A body shell. The French term coque has the same meaning, but is only used in the form monocoque, meaning single-shell.

shell bearing Plain bearing formed from two interlocking and abutting thin-walled semi-circular cusps. See Figure C.10.

shell structure Vehicle structure consisting predominantly of stress-carrying panels. See also ***monocoque***; ***unitary construction***.

shift collar Splined sliding collar by which gears are selected in a ***constant mesh gearbox/transmission***. See Figure S.13.

shift fork (UK: selector fork) Forked member for moving a sliding pinion into and out of engagement in a ***change-speed transmission***. See Figure S.1.

shift lever See ***gearshift lever***. See Figure S.1.

shift rail The rod or "rail" on which the shift fork of a change-speed gearcase travels. See Figure S.1.

shift valve Valve that actuates the automatic gear change in an ***automatic transmission*** following the signal from the throttle and governor.

shifter rod Rod which communicates the movement of a gear-shift lever with a remotely mounted transmission.

shimmy Low-amplitude mechanical vibration, particularly as an imbalance fault of steered wheels. See also ***lateral runout***.

shock A suspension ***shock absorber***. (Informal)

shock absorber Mechanism for damping vibration in a sprung system, such as a vehicle suspension. Also ***shock damper*** or ***damper***. A telescopic ***hydraulic damper***. See Figure S.4.

shocker A ***shock absorber***. (Slang)

shoe See ***brake shoe***. See Figure D.8.

shoe brake A ***drum brake*** with shoes. See Figure D.8.

shooting brake See ***estate car***; ***station wagon***. (UK archaic)

short engine (1) Incomplete engine, usually lacking cylinder head and ancillaries. (2) A reconditioned engine as in (1).

shoulder (1) The outermost edges of a tire's ***tread***. (2) An abrupt increase in diameter of a bolt, stud, or other fastener. See Figures B.3 and R.2.

shoulder harness Belt restraint system in which an anchored length of webbing passes over one shoulder of an occupant and is fastened at waist height. May be used in conjunction with a ***seat belt***.

shoulder rib Rib or squared-off rim at the outer edge of a tire tread. See also ***tread shoulder***.

shoulder wear Tire wear pattern where wear is greatest at the tire's ***shoulder***.

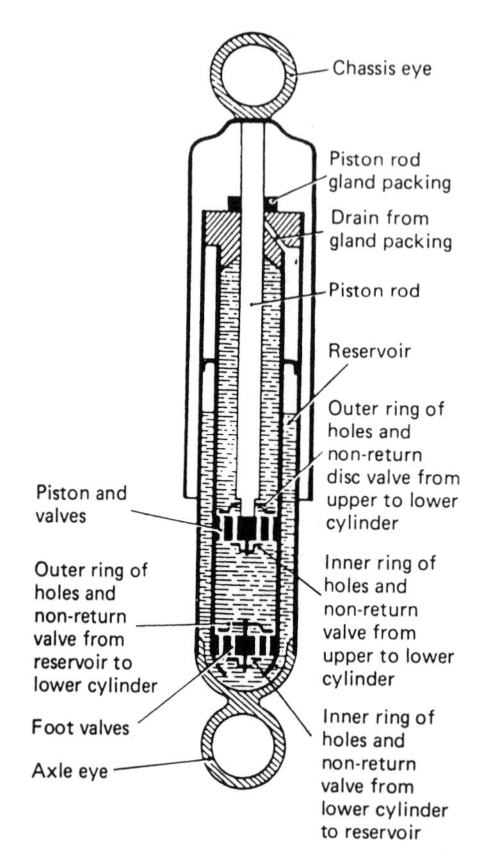

Figure S.4 A telescopic shock absorber.

showman's vehicle Vehicle for the transportation of fairground and circus equipment and artefacts, etc. Exempt from normal regulations in some countries. See also ***heavy haulage***.

shut line Visual line of styling significance around closed doors, hatches and other hinged features.

shutter Mechanical device consisting of movable vanes for restricting the flow of air, as for example air flow to cool an engine. See also ***radiator blind***.

shuttle dumper A ***dump truck*** for use on narrow sites such as tunnels and cuttings, capable of driving in forward and reverse, and equipped with a revolving driver's seat and controls accessible in both directions of operation.

shuttle valve A three-way valve with two input ports and one output port, which selects the higher pressure of the two inputs, and directs this to the output port.

siamesed Abnormally joined or paired: (1) Of cylinder bores, joined metal-to-metal to the full depth of the cylinder, thus precluding the flow of cooling water between them. (2) Of exhaust pipes, parallel pipes joined along part or all of their length. (3) Of valve ports, arranged so that two adjacent valves are served by one port.

side branch resonator Acoustic resonator fitted as a dead-end branch to a relevant tract, such as inlet or exhaust, there being no flow through the resonator. See also ***Helmholz resonator***; ***quarter-wavelength resonator***.

side draft carburetor Carburetor with horizontal ***barrel***.

side dumper Dump truck capable of discharging to the side rather than (or as well as) to the rear.

side guard Protective structure built along sides of goods vehicles forward of rear wheels to prevent intrusion of other vehicles on impact, and to minimise injury to pedestrians and cyclists. See also ***rear underrun bumpers***.

side impact bar See ***side intrusion bar***.

side intrusion bar Reinforcing beams, usually set within the door structure of a vehicle, to reduce intrusion in side impact.

side lamp Low-intensity lamp mounted on each forward corner of a vehicle to indicate presence from front. See also ***parking lamp***.

side light The light emitted by a side lamp (and not the item itself).

side marker lamp Lamp mounted on the side of a long vehicle or vehicle combination particularly to indicate presence to other vehicles.

side panel Raised panel at either side of an open truck's platform, to retain the load. See also ***dropside flat***. Also ***sideboard***.

side splitter Horizontal end-plate along lower bodywork of a vehicle, to control airflow.

side stand Retracting leg to support a ***motorcycle*** in an upright position when parked.

side valve (US: L head, T head) Of an engine, having the valves and valve gear to the side of the cylinder rather than above it, and normally mounted in the cylinder block. See also ***overhead valve***. See Figure S.5.

sideboard See ***side panel***.

sidewall (1) The side of a ***tire***, between the ***tread*** and ***bead***. (2) Side panel, particularly of a truck or box-van.

silencer (US: muffler) Box or chamber within an engine ***exhaust system*** for reducing exhaust noise. See also ***exhaust pipe***. See Figure E.2.

Silentbloc bearing Proprietary rubber/metal anti-vibration mounting.

silicones Synthetic polymeric lubricants added to mineral oils to inhibit foaming.

sill Horizontal member forming the structural side of a ***floor pan***, and onto which a door or ***tailgate*** closes. Also ***cill***. See Figure B.4.

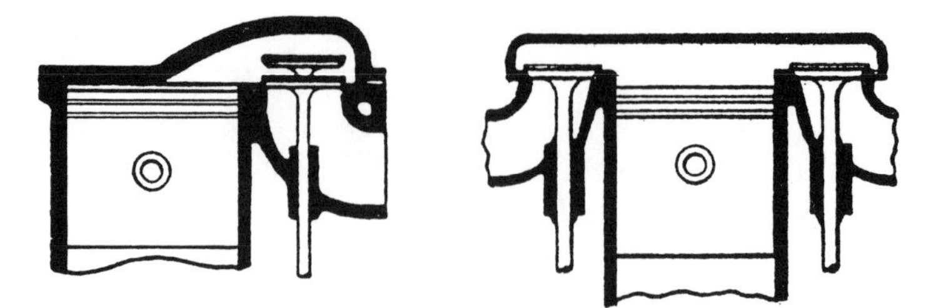

Figure S.5 Side valve configurations: A Ricardo-style turbulent L-head of the mid-1930s (left) and the much earlier T-head (right).

single acting engine Engine in which the working pressure of combustion gases is applied to one side of the piston only, as is normal practice in automotive engines.

single anchor brake ***Drum brake*** in which the ***leading*** and ***trailing shoes*** both pivot about one fixed stud or anchor.

single axle weight Total weight transmitted to the road by wheels whose axes are 40 inches or less apart for purposes of US legislation.

single beam headlamp Headlamp with a single filament and therefore able to provide main (upper) or dipped (lower) beam but not both. See ***dual beam headlamp***.

single leaf spring Single element suspension spring that operates as a flexible beam, supporting the vehicle axle at mid-section. See also ***semi-elliptical spring***.

single overhead camshaft Valve train camshaft operating both intake and exhaust valves. Also ***sohc***. See also ***twin camshaft***.

single pivot steering Steering in which a beam axle is pivoted at its mid-point. Rare except on small agricultural and horse-drawn vehicles. See also ***Ackermann steering***; ***double-pivot steering***.

single plane crankshaft Crankshaft in which cranks are in line or disposed at 180 degrees to one another. See Figure E.1.

single plate clutch Clutch with one ***driven plate***. See Figure C.4.

single point injection Gasoline engine ***fuel injection system*** in which fuel is sprayed into the intake ***manifold***, in a similar way to a ***carburetor***. Also ***SPI***.

single-reduction axle ***Live axle*** in which the reduction in rotational speed between the ***propeller shaft (drive shaft)*** and the final drive is in one stage, normally within the ***differential casing***. See Figure D.5.

single track vehicle Vehicle, such as a bicycle or motorcycle, that leaves only one track.

sipe Usually shallow groove or channel in a ***tire tread pattern*** that may serve to reduce tire noise, dissipate heat or increase flexibility of the tread pattern.

sisters Twin clamping but adjustable buckles for securing webbing, as of a ***seat belt***.

sizzle Noise of ***tire tread*** on road surface characterized by a soft "frying" sound.

skeletal Commercial vehicle, ***semi-trailer*** or ***full trailer*** chassis, on which sturdy cross members or bolsters are mounted to support a ***container*** or other boxed load.

skeleton wheel Fabricated metal wheel for off-highway and particularly agricultural tractor use, and generally equipped with soil-gripping lugs attached to annular metal outer rims. See also ***rim wheel***.

skid (1) Motion of a vehicle when the wheels are locked or partially locked, or when the vehicle moves sideways as in a side slip. A ***wheel skid***. (2) Flat wooden base to facilitate lifting or loading of heavy items. (3) A pallet. (US informal)

Skid Number Arbitrary scale for expressing the frictional resistance of a pavement, initiated by ASTM.

skid plate Any deflector plate under a vehicle that provides protection, as to the ***oil pan (sump)*** of an engine. (US informal)

skid steer Steering of a vehicle, particularly an off-highway and earthmoving vehicle, by braking of the wheels on one side and driving on the other so that direction is changed by skidding around. The term is applied particularly to pneumatic tired vehicles.

skin friction Aerodynamic drag resulting from the movement of air around the surface of a vehicle.

skip Demountable hopper, usually for refuse or builder's waste. (UK)

skip loader Vehicle for the loading, unloading and conveyance of ***skips***.

skipshift Gear shifting (gear changing) between two non-contiguous gears, as from second to fourth.

skirt (1) The wall of a piston below the piston crown that spreads side-thrust loads. See Figure P.2. (2) Any downward planar extension of a vehicle body for the control of air flow or the reduction of road soiling.

skylight Transparent screen in the ***roof panel*** of a vehicle.

slack adjuster (1) Device that keeps the pushrod and cam lever of an air brake at approximately right angles to provide maximum effort despite wear of linings. (2) An adjustable member which transmits brake application force and permits compensations for lining wear.

slap (1) Smacking noise produced when a ***tire*** traverses road seams or other surface irregularities. (2) Piston slap.

slave cylinder Cylinder and piston which, under hydraulic or pneumatic pressure from a ***master cylinder***, actuates mechanical components as for example a ***brake*** or ***clutch slave cylinder***.

sleeper cab Commercial vehicle cab providing sleeping accommodation for a driver. See also ***top sleeper***.

sleeper pod Sleeping compartment, sometimes not integral, mounted above the cab of a commercial vehicle. Also ***top sleeper***.

sleeping policeman Shallow hump on a road surface, built across the traffic flow, to deter driving at excessive speed. (UK informal) Also ***speed bump***. (US)

sleeve (1) **(UK: cylinder liner)** Hard metal tubular lining to an engine ***cylinder***. See Figures A.2 and W.4. (2) Any tubular component, usually finished to close tolerances, that fits within an aperture or onto a shaft, but particularly one that is long in relation to its diameter, whereas a ***bush*** is usually short. (3) A ***sleeve valve***.

sleeve connector A tubular electrical connector, often of the ***bullet type***. (Obsolete)

sleeve valve Mechanically operated tubular sleeve or liner, with apertures, located between ***piston*** and ***cylinder wall*** of an engine, and serving as an ***intake valve*** and ***exhaust valve*** through its rotary and/or oscillatory motion which brings the apertures in the sleeve in conjunction with the appropriate ports.

sleeve-valve engine Engine employing a sleeve valve controlled induction and exhaust system. Currently rare in automotive practice.

slew To rotate precisely about a vertical axis, as of the body of a ***mobile crane*** or ***excavator***.

slewing ring The bearing on which the rotating part of plant, such as a crane or excavator, rotates or slews. Also ***slew ring***. See Figure C.12.

slide operated carburetor Carburetor in which the volume flow of air is regulated by a simple slide valve in the inlet tract. Common on motorcycles and light machinery.

slider Commercial vehicle tractor unit with facility for fore-and-aft adjustment of the ***fifth-wheel*** to suit different lengths of ***semi-trailer***. (Informal)

sliding-block joint See ***de Dion joint***.

sliding dog (1) Internally splined collar that slides along the mainshaft of a change-speed transmission under the action of a ***selector fork*** and so ensures positive and synchronous engagement of mainshaft gear to mainshaft. Sometimes called ***synchronizer*** or ***synchromesh sliding sleeve***. (2) Any form of axially mobile dog clutch.

sliding fifth wheel System of positive but adjustable fore-and-aft location of a ***fifth-wheel*** of a commercial vehicle tractor unit to accommodate different lengths of ***kingpin*** position and axle loading of ***semi-trailers***.

sliding joint A shaft joint of coupling that allows axial movement between the joined shafts. A ***plunging joint***. See also ***pot joint***.

sliding mesh gearbox Gearbox in which the change of gear ratios is accomplished by sliding pinions along a splined main shaft to engagement with fixed wheels on a ***layshaft*** or ***countershaft***. See also ***constant mesh gearbox***; ***crash gearbox***; ***selector fork***.

sliding spline Splined joint in a shaft capable of accommodating axial movement while transmitting torque. Also ***slip-spline***. See Figure H.2.

sliding tandem Two-axle commercial vehicle undercarriage with facility for longitudinal relocation to obtain a favorable load distribution.

slinger (1) Disc used for imparting radial momentum to a liquid, and particularly a lubricating oil, to prevent impingement on a shaft seal. (2) Device used to throw oil from a sump or oil pan onto surfaces requiring lubrication. See also ***oil flinger***.

slip angle Angle between plane of wheel and direction of travel of ***center of tire contact***. Also ***distortion angle***. See also ***toe-in***; ***toe-out***. See Figure U.1.

slip splines See ***sliding spline***.

slip yoke Yoke of a Hooke's joint in a vehicle transmission, and particularly a Hotchkiss drive, that accommodates axial end-movement by way of ***sliding splines*** or ***slip splines***. See Figure H.1.

slippage Sliding movement between ***tire***, or track of a ***track-laying vehicle***, and the ground.

slipper block Robust seating that supports one end of a ***leaf spring*** and allows sliding fore and aft movement as spring deflects. See also ***slipper spring***. See Figure T.1.

slipper pretensioner A usually curved and usually self-adjusting member that bears on a roller chain and so takes-up slack due to fitting tolerances and wear.

slipper skirt piston Piston with skirt cut away beneath ***piston pin*** axis to provide ***crankshaft*** clearance and reduce friction. Also ***partial skirt piston***.

slipper spring Leaf spring of which only one end is fixed to the chassis, the other end resting on a slipper block.

slipstreaming Using the wake of a preceding vehicle to reduce air resistance of the following vehicle.

sloper Single cylinder or parallel engine of which the cylinder axes are inclined to the vertical. (Informal)

slotted piston (US: split skirt piston) Piston in which the ***skirt*** is slotted to counteract effects of thermal distortion.

slow running jet ***Carburetor*** jet that compensates for the natural reduction in mixture strength at low engine speed by supplying excess fuel. Also ***idling jet***. See also ***idle system***.

sludge Thick deposit formed from ***lubricant*** stiffened with products of combustion and partial combustion such as unburned ***hydrocarbons***, carbon particles, oxides and aldehydes.

sludge trap Cavity in the lubrication system of an engine, but particularly a ***dry-sump engine***, to prevent ***sludge*** being circulated in the flow of lubricating oil.

smoke Visible emission consisting primarily of suspended ***particulates*** or vapor.

small end The piston pin end of a ***connecting rod***. (UK informal) See also ***little end***; ***piston pin end***. See Figure C.10.

smoke opacimeter Optical instrument for measuring the opacity of exhaust gases.

smoke tunnel Wind tunnel in which streams of smoke enable the air flow over a vehicle or component to be visualized.

smokemeter Meter for measuring smoke content of ***exhaust emissions***.

snap connector (1) An electrical connector with positive spring engagement between male and female parts. A ***bullet connector*** or ***sleeve connector***. (2) Any form of connector, mechanical or otherwise, with location by positive sprung action.

snow chain Link chain attached to ***tire*** to improve traction in snow.

snowmobile Vehicle designed primarily for off-highway mobility on snow.

snub Act of retarding a motor vehicle between two speed values by use of the brake system.

snub braking Short-duration braking to reduce the speed of a vehicle to a lower level.

snubber (UK: bump stop) Compressible rubber, spring, or other device to prevent excessive suspension travel, or to act as a damper on ***rebound***.

soak In the context of heat, the stabilizing of temperature throughout a body, such as an ***engine*** that has been running for short period, whereby the parts more distant from the ***cylinders*** reach their working temperature.

soak time Time required for an ***engine*** to reach a stable temperature throughout all components, as for example prior to cold start emission tests.

sodium cooled valve Engine ***exhaust valve*** cooled by the agency of sodium within a hollow ***valve stem***. Movement of the sodium, which liquefies at engine running temperature, conveys heat from the valve head to the stem, where it is conducted to the valve guide.

sodium-sulfur cell Battery in which two liquid electrodes are separated by an ionically conductive ceramic electrolyte, with a molten sulfur anode and molten sulfur-sodium polysulfide as the cathode.

soft plug A hot grade of ***spark plug***.

soft top (1) The fabric, usually folding, top of a ***convertible*** vehicle. (2) Vehicle fitted with a soft top.

solenoid See ***solenoid switch***.

solenoid switch Electromagnetically activated electrical switch, often used where a high current circuit, such as a starter motor circuit, is brought into operation by a low current switch. See ***starter solenoid***.

solid-axle suspension Suspension in which wheels are mounted at either side of a rigid ***beam axle***, so that any vertical movement of one wheel is transmitted to the opposite wheel. A ***non-independent suspension***.

solid injection Fuel injection in which the fuel charge is injected under its own hydraulic pressure and not as part of an airborne spray. Also ***airless injection***.

solvent Liquid that is capable of dissolving another material by chemical action.

sonic throttling Increasing the flow velocity of a gas or vapor in a pipe, tube or orifice to sonic speeds by throttling, as applied to ***carburetion*** or engine induction air flow.

sound energy density Acoustic energy per unit volume of material.

sound intensity Rate of flow of sound energy in specified direction per unit area.

space frame Three-dimensional structural framework that serves as a chassis and anchorage for body panels as an alternative to a monocoque or conventional chassis structure, particularly in racing cars. See Figure S.6.

spark advance See ***advanced ignition***.

spark gap (1) Gap between the ***electrodes*** of a ***spark plug***. See Figure S.7. (2) Safety gap in a ***magneto***.

spark ignition System that uses an electric spark to ignite a fuel/air mixture.

spark ignition engine (1) Engine running on the ***Otto cycle***. (2) Any form of reciprocating engine in which combustion is initiated by a spark.

spark knock Detonation within an engine cylinder caused by excessively ***advanced ignition timing*** rather than to surface ignition. See also ***detonation***; ***knock***; ***pre-ignition***.

spark plug Insulated plug that supports the electrodes between which the high voltage spark passes to initiate ignition in a spark ignition engine. Also ***sparking plug***. (Mainly UK) See Figure S.7.

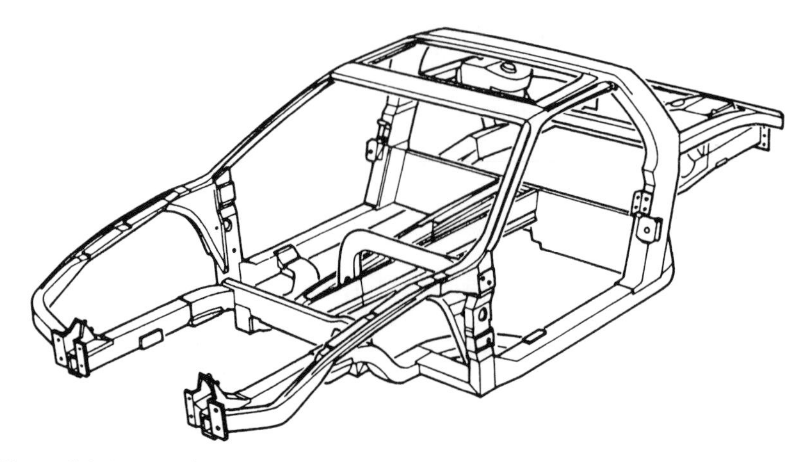

Figure S.6 A pressed-steel space frame for a sports car (TI Automotive).

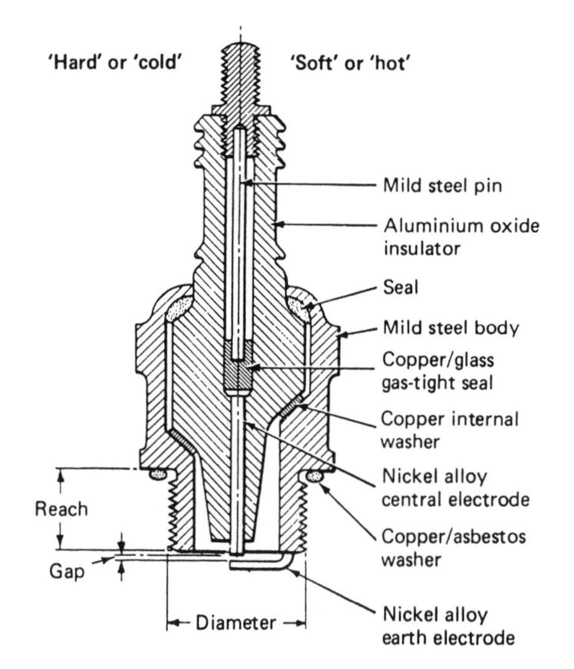

Figure S.7 Sectional views of a "cold" spark plug (left) and a "hot" plug (right).

specific energy Of an electrical battery or cell, the electrical capacity per unit weight, usually expressed as Watt-hours per kilogram.

specific fuel consumption Fuel consumed per unit of power output, usually expressed as mass or volume per unit power per unit time, for example litres per kilowatt hour, or pounds per brake horsepower hour. Also ***SFC***.

specific power Of an electrical battery or cell, the power available per unit weight, usually expressed in Watts per kilogram. Power will normally vary with state of charge and rate of discharge.

speed bump See ***sleeping policeman***.

spider (1) Any component with any number of legs or shafts radiating from a central hub. In nature the spider has eight legs. (2) Cruciform wheel-wrench having different sized box-heads on each leg. (3) In certain types of pressed wheel, the component that secures the wheel to the hub. (4) The cruciform linking member of a Cardan or Hooke's joint. See Figure H.1. (5) High-performance sports car. (German and Italian)

spike stop Stopping or deceleration of a vehicle brought about by sudden application of a heavy brake pedal force.

spill port Hole through which excess fuel is returned to the fuel system in a diesel engine ***injector pump*** at the end of injection.

spin To crank a motor for starting. (US informal)

spin axis Axis about which a wheel spins.

spin turn To turn a vehicle, and particularly a ***track-laying vehicle*** or pneumatically tired off-highway vehicle, by applying forward drive to one side and reverse or full braking to the other so that the vehicle effectively turns about its vertical axis. See also ***skid-steer***.

spin velocity Angular velocity of a wheel about its ***spin axis***.

spine frame Structural backbone, particularly of a motorcycle, consisting of one torsionally stiff member, usually a welded pressing. Also **spineframe**.

spindle See ***kingpin***.

spindle arm (UK: steering arm) Lever attached to the ***stub axle*** assembly of a steered wheel whereby the movement of the ***drag link*** is converted to the steering movement of the steered wheel about its steering axis. Occasionally the arm that links the ***track rod*** (tie rod) to the ***steering knuckle*** (stub axle assembly).

spine-back Narrow central chassis to which running gear and bodywork is attached, mainly on heavy-duty commercial vehicles and ***semi-trailers***. Also ***spine chassis***.

spine chassis See ***spine-back***.

spinner (1) Clutch driven disc. (UK informal) (2) Knock-on wheel nut. (UK informal)

spiral bevel ***Crown wheel*** in which the teeth radiate as part of a geometric spiral, commonly used in final drives.

spiral bevel axle A drive axle fitted with a ***spiral bevel differential***.

spiral bevel differential Differential with a spirally cut ***crown wheel*** and pinion.

splash guard (UK: mud flap) Flexible deflecting shield mounted behind roadwheels of a vehicle to control spray.

spline Parallel key around a shaft, in appearance rather like the teeth of a spur gear, that slides into a suitably keyed female component, and thereby forms a joint by which torque and motion can be transmitted. Splined joints may be fixed, where there is no relative axial motion, or ***sliding splines***. See Figure H.1.

spline grunt Noise emanating from ***sliding splines*** or slip splines when under load, particularly at lower speeds.

split axle casing Axle casing or axle housing (US) made from two halves bolted together, as opposed to a ***banjo axle*** casing, which is normally in one piece.

split braking system System of braking consisting of two or more separate systems such that failure of one would leave the other as an effective system for braking the vehicle. Normally the primary means of braking, for example a pedal, would be common to both or all systems.

split crankcase Crankcase comprising two or more parts, usually divided in the plane of the crankshaft to facilitate assembly. See also ***monobloc construction***; ***unitary construction***.

split crankshaft Crankshaft assembled from individual components rather than forged or cast as a unit.

split cycle engine Engine in which the normal cycle of operation is apportioned between two separate cylinders, or in which there are more than four separate cycle activities.

split service brake See ***split braking system***.

split single ***Two-stroke engine*** in which two ***pistons*** and ***cylinders*** share the same combustion chamber. Also called ***U-cylinder***.

split skirt piston (UK: slotted piston) Piston in which the skirt is slotted to reduce thermal distortion. See also ***T-slot piston***.

split system See ***split braking system***. Also ***divided system***. (Mainly UK obsolescent)

split wheel Wheel which is split, usually down the centerline, with the two parts bolted together. Also ***divided wheel***. See also ***two-piece wheel***.

splitter transmission Manual change-speed transmission with additional gearing that interposes an extra output ratio between the basic ratios, thus increasing the number of available speeds. For example, a lower and three intermediate speeds might be added to a nominally four-speed transmission. The splitter is usually engaged by an automatic ***pre-selector***. Also ***splitter box***. See also ***auxiliary gear-box***; ***range change***.

spoiler Any transverse aerodynamic device attached to or built onto a vehicle to modify air flow. Originally an extensible vertical flap to reduce or "spoil" the lift of the wing of an aircraft on landing. See also ***air dam***; ***apron***; ***Gurney flap***.

spoke Linear bracing between ***hub*** and ***wheel rim***, of wire, cast metal, or in earlier wheels, of wood.

spot lamp Front-mounted narrow beam lamp to be aimed at the will of the driver, normally restricted to use when vehicle is stationary. Illegal in UK where lamp, usually fitted to bumper as an accessory must not have provision for aiming from within vehicle. Also ***long range lamp***.

spot-type disc brake A conventional ***caliper disc brake***.

sprag clutch Type of ***one-way clutch*** or ***freewheel*** using spring loaded locking "cams" between inner and outer races.

spray holes Apertures in an ***injector*** through which fuel is introduced into an engine or its inlet tract. See Figure Q.1.

spread tandem axle Widely spaced ***tandem axle*** arrangement of a heavy vehicle, usually employed to circumvent axle weight restrictions applying to conventional close-coupled tandem arrangements.

spring aid See ***bump stop***.

spring base Transverse distance between the points of action of the springs of a vehicle ***suspension***. See Figure T.6.

spring brake Pneumatic parking or ***secondary brake*** in which a helical spring is compressed by action of air pressure on a piston when the brake is not required, the pressure then being reduced to apply the brake. Pressure within a second chamber causes the spring brake to assist the ***service brake*** as a secondary brake. See also ***fail-safe spring brake***.

spring flange Removable spring steel flange for retaining ***tire*** on ***two-piece wheel rim***. See also ***bead flange***.

spring hanger Bracket on a vehicle ***chassis*** to which a ***leaf spring*** end is attached.

spring liner Soft material set between the leaves of a ***leaf spring***.

spring pocket Recess in a mounting bracket or other chassis member to locate a suspension coil spring. Also ***spring seat***. See also ***spring tower***.

spring rate Load per unit deflection, or sometimes deflection per unit load, of a spring or of a sprung system.

spring seat See ***spring pocket***.

spring tower Raised disc to locate a suspension ***coil spring***.

spring wind-up Deflection of a suspension spring, particularly a ***leaf spring***, resulting from torsional loads of braking or acceleration. See also ***axle wind-up***.

sprocket drive A toothed driving wheel, particularly the drive to the ***track*** of a ***crawler***.

sprung mass (1) The body and chassis of a vehicle that rides on the suspension springs. (2) The mass of suspended body and chassis, sometimes including half of the suspension mass.

sprung weight See ***sprung mass***.

spur differential Differential employing spur gear wheels rather than bevel wheels, as in the more common ***bevel differential***.

squab The upholstered backrest of a ***seat***. (UK)

squab panel Nominally vertical transverse panel which supports the rear seat ***squabs*** and separates the passenger compartment from the luggage space in a passenger car.

square engine Engine in which the cylinder bore diameter is equal to the piston stroke. See also ***oversquare engine***; ***undersquare***.

square four Four-cylinder engine in which each cylinder axis forms one corner of a square, the engine being equivalent to two parallel vertical twin engines.

squat Tendency of a vehicle body to be depressed toward the ground as a result of torque reaction to acceleration. See also ***dive***.

squeal (1) Noise of high-frequency vibration from a brake assembly. (2) Airborne tire noise resulting from ***slip*** or ***skidding***.

squish (1) The squeezing of air-fuel mixture in an engine's combustion chamber away from the end-gas region at the top of the compression stroke. (2) Area of an engine combustion chamber with minimal clearance between ***piston crown*** and ***cylinder head*** at ***top dead center***.

squish lip Bowl-in-piston ***combustion chamber*** featuring squish with a re-entrant bowl and hard-edged entrance from bowl to crown, originally developed by Perkins.

squish motor Engine in which ***squish*** is a prominent feature of the ***combustion chamber*** configuration.

stability Tendency of a vehicle to remain in a steady state under the influence of perturbing forces, or to return to that state when momentarily disturbed from it.

stabilizer bar (UK: anti-roll bar) Torsion bar coupling nearside and offside wheel suspensions of an independent suspension system, to minimize body roll. See Figure A.3.

stabilizer ply Tire ply extending only from ***shoulder*** to shoulder of a ***tire***. The tire ***plies*** together form a belt. See also ***cap ply***; ***carcass ply***.

stack height Of a ***leaf spring*** assembly, the total depth over the individual leaves at the spring's mid-section. See Figure L.2.

staggered V-engine Engine with cylinders set in two banks with their planes of axes at an acute or narrow V angle, the axes of the individual cylinders in each bank staggered to obviate interference of the bores, originally devised (by Lancia) to provide a compact engine geometry.

stainless steel Range of steels of varying specialist properties and compositions, of bright, silvery appearance, containing chromium and other alloying elements, notably ***nickel***. Used decoratively and for engineering applications where high corrosion-resistance is important, but rarely for primary structure.

stall (1) Inadvertent stopping of an engine, usually due to sudden increase in load without a commensurate increase in engine speed. (2) Engine rotational speed when driving through an ***automatic transmission*** with the vehicle stationary.

stanchion Sturdy vertical member, as for restraining a load.

stand pipe Vertical air-pipe with fitting at upper end for attachment of trailer air pressure hose to air brake system of tractor unit. See also ***quick-detachable coupling***.

star wheel An adjusting nut with indexing arms for positive and accurate rotation, as on a brake or clutch.

start-up groan Low-frequency noise, often transmission shaft generated. Also ***G-string***.

starter motor High torque electric motor for starting an engine, normally by a high ratio geared drive to the ***flywheel ring gear***. Also ***cranking motor***. (US informal) See also ***Bendix drive***; ***ring gear***.

starter solenoid A solenoid switch for operating a ***starter motor***.

starting handle Cranked handle for manually starting an engine.

static seal Seal to retain pressure or prevent loss of fluid between two fixed surfaces or components. See also ***dynamic seal***; ***gasket***.

static tire deflection Effective decrease in wheel radius from ***center of tire contact*** to ***spin axis***.

static toe Difference in distance at extreme points of ***tire tread*** between front and rear of a pair of wheels with the vehicle stationary. See also ***toe-in***; ***toe-out***.

station wagon (UK: estate car) Passenger car with extended constant height body fitted with tailgate or rear doors to facilitate access and provide stowage for bulky items. Also ***wagon***; ***shooting-brake***. (UK archaic)

stator The non-moving parts of a rotating system, as for example the static blades of a torque converter or the static winding of a ***dynamo*** or ***alternator***. Note that in some ***torque converters*** the stator may rotate in free-wheeling mode.

steam reformation Process by which hydrogen can be separated from natural gas or other hydrocarbons. The steam chemically reacts with the ***hydrocarbon***. The hydrogen comes from the steam as well as from the hydrocarbon. Also ***steam reforming***.

steel A malleable metal consisting mainly of iron and carbon, produced with a wide range of properties but most widely as mild steel.

steerable wheel See ***steered wheel***.

steered wheel Road wheel that responds to applied steering input and therefore influences the path taken by the vehicle.

steering angle (UK: angle of lock) Angle between the projection of the longitudinal axis of a vehicle and plane of ***steered wheel***. Also ***steer angle***. (US)

steering arm (1) Lever that imparts the steering action to the ***steering knuckle*** or ***stub axle*** assembly, as from a ***tie rod*** or ***track rod***. Also ***spindle arm***, though this term is often used for the lever that imparts steering action from drag link to steering knuckle. (US) (2) An intermediate arm in a steering system. See Figure S.8.

steering axis The axis about which a ***steered wheel*** assembly pivots when steered. Also ***kingpin axis***. See also ***caster***. See Figure C.2.

steering box (US: steering gear) Gearbox in which the rotary movement of the ***steering column*** is converted to the angular motion of the ***drop arm*** or ***Pitman arm***. See also ***Gemmer***; ***Marles***; ***recirculating ball steering***; ***worm and nut steering***; ***worm and peg steering***; ***worm and sector steering***. See Figures S.8, W.6 and W.7.

steering column Shaft that transmits the rotation of the steering wheel to a steering box or rack and pinion. Also ***steering post***. (US obsolete) See Figure P.5.

steering column gear change (US: steering column gear shift) Gear shift mechanism in which the shift lever is remotely mounted on the steering column, operating the ***selector forks*** by means of a system of rods and levers. Also ***column change***.

steering damper Device for damping vibrations or shock loads in the ***steering system***.

steering feel Subjective quality of the steering response of a vehicle. See also ***road feel***.

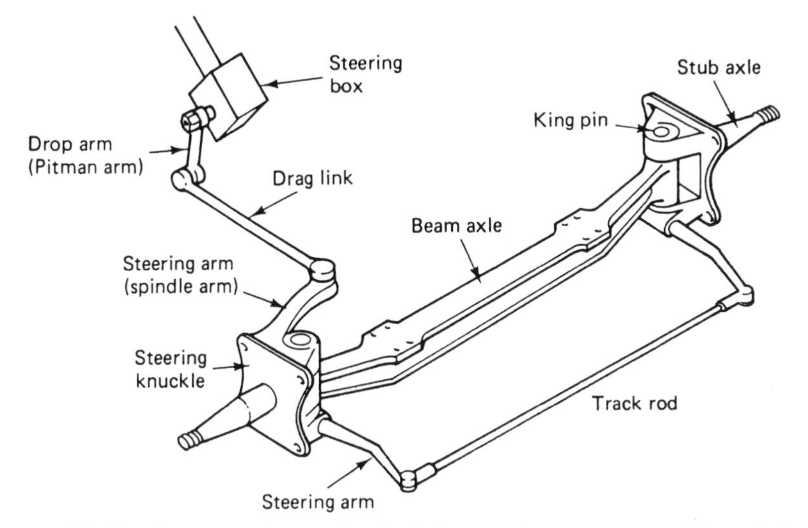

Figure S.8 Steering nomenclature of a simple beam axle arrangement. Many of the items have alternative names.

steering gear (1) The components by which a vehicle is steered. (UK) (2) **(UK: steering box)** Gearcase in which the rotary movement of the ***steering column*** is converted to the angular motion of the ***drop arm*** or ***Pitman arm***.

steering geometry The geometric arrangement of the components and linkages of a steering system, and the numerical value of the lengths and angles thereof. See also ***suspension geometry***.

steering head Tube or socket at the forward end of a motorcycle or bicycle frame which provides the pivot bearing for the ***front forks***.

steering knuckle Assembly or component comprising ***stub axle*** and ***steering arm***.

steering lock (1) Angular travel of steered wheel from straight ahead to ***full lock***, the extreme angle in either direction. (2) Security locking device that disables steering.

steering post Shaft that transmits the rotation of the ***steering wheel*** to a ***steering box*** or ***rack and pinion***. (US obsolete)

steering ratio Ratio of angular movement of ***steering wheel*** to change of ***steering angle*** of the ***steered wheels***.

steering side tube See ***drag link***.

steering spindle A steered ***stub axle***. See Figure S.8.

steering system Mechanism or means whereby the direction of a vehicle is controlled. See Figure S.8.

steering tie-rod Rod that connects the ***steering arms*** (knuckle arms) of one steered wheel to another. (US) ***Track rod***. (UK) See Figure S.8.

steering wheel The wheel by which the driver controls the direction of a vehicle. See Figure P.5.

steering worm Worm for winding steering cable or chain on a steered ***beam axle***, as on a steam traction engine.

stellite Proprietary hard metal alloy noted for hardness and wear resistance at high temperature and used for coating exhaust ***valve seats***.

step-frame trailer Trailer, and particularly a ***semi-trailer***, of which the forward part of the platform is raised by a step to clear the tractor chassis, the after part being lower to facilitate loading. See also ***drop frame***.

stepped piston engine Engine with pistons stepped to give two diameters, smaller and larger diameter parts normally fulfilling different functions, such as pumping and power. See Figure S.9.

stepped reflector Reflector, particularly of a ***headlamp***, consisting of parabolic sections of different focal length. See also ***homofocal reflector***.

stepless transmission Transmission with continuously variable ratio of output to input speeds, usually, though not necessarily, automatic. See ***continuously variable transmission***; ***infinitely variable transmission***.

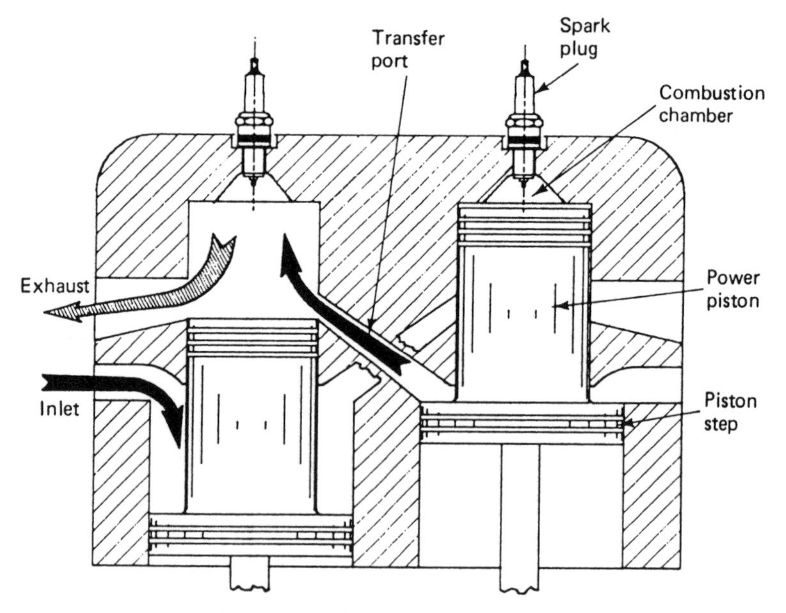

Figure S.9 Cylinder and head of the Hooper stepped piston engine.

stick-shift Manual transmission, with shift lever set on floorpan or transmission tunnel. (Mainly US usage)

stick-slip Irregular, noisy motion of sliding parts due to ***boundary lubrication*** failure and high static friction.

Stirling engine An ***external combustion engine*** working on a closed regenerative thermodynamic cycle, with cyclic compression and expansion of a gas in which flow is controlled by volume changes, usually between two co-axial pistons. The practical Stirling engine does not work on the idealized Stirling cycle.

stoichiometric ratio Ratio, usually of mass, between air and flammable gas or vapor, at which complete combustion or chemical combination takes place. Previously called chemically correct mixture strength.

stop lamp (UK: brake lamp) Rear signal lamp which illuminates when ***brake*** is applied.

stoplight drag Clutch slipping on a revving engine to gain high acceleration at the expense of clutch wear and transmission wind-up. (US informal) Also ***jack-rabbit start***.

storage battery See ***battery***.

straddle carrier Goods or industrial vehicle which straddles its load for lifting.

straddle trailer Framework trailer which is driven over the cargo, which it lifts by an integrated lifting and suspension system.

straight engine An ***in-line*** engine. The term is often used in conjunction with the number of cylinders, a straight eight being an in-line eight-cylinder engine. (Informal)

straight oil Mineral oil containing no ***additives***.

straight truck (UK: rigid) Truck with body, cab and engine mounted on the same chassis. A non-articulated commercial vehicle.

strainer A filter, particularly a coarse ***full-flow filter***.

strangler Flap that restricts the flow of air into an engine ***induction system***. A ***choke***. See Figure C.3.

strap drive Tangential steel springs by which a ***clutch pressure plate*** is attached to the rotating clutch cover, thereby providing a degree of axial and rotational flexibility.

stratified charge Reciprocating engine combustion system in which combustion is initiated in a layer of relatively fuel-rich mixture, and then spreads to a much weaker mixture elsewhere in the combustion chamber to increase fuel economy.

streamlined Shaped to minimise air resistance, or appearing to be so. See also ***drag***; ***drag coefficient***; ***profile drag***.

streetcar Passenger-carrying rail vehicle, usually electrically propelled, that runs on tracks laid within a street system. Also ***tram***. (UK informal)

Stribeck Curve Plot of friction against ZN/P where Z is viscosity, N bearing speed and P bearing pressure, from which transition from ***boundary*** to ***full fluid film lubrication*** can be observed.

striking fork Attachment that moves a ***selector rod*** in a ***change-speed gearbox***. See Figure S.1.

strobe A stroboscope, as for timing an engine ignition system. (Slang)

stroke Total axial movement of a ***piston*** in its ***cylinder bore***, equivalent to the diameter of the circle described by the ***crankpin axis***.

stroke-bore ratio Ratio of stroke to cylinder diameter in an engine. See also ***oversquare engine***; ***square engine***; ***undersquare engine***.

stroking Increasing the swept volume of an engine by increasing the stroke. (Informal)

stub axle Short cantilevered ***axle*** on which a wheel is mounted. See Figure S.8.

studio buck Styling model of a vehicle in clay, wood or other material, usually without detail. See also ***clay buck***.

sub-frame Usually removable chassis sub-structure on which may be mounted, for example, a suspension and steering system. Also ***subframe***.

sub-harness A subsidiary part of an electrical ***harness*** which can be disconnected from the main harness.

submarining Form of malfunction of a vehicle ***restraint system*** in ***impact*** in which the occupant slides forward under the ***seat belt***.

suction manifold See ***intake manifold***.

suction stroke The ***induction stroke*** or ***intake stroke***.

sulphonates (US: sulfonates) A class of chemicals, some of which have been used as oil and fuel detergent additives.

sulphur oxides (US: sulfur oxides) Generic term for the various oxides of sulphur, particularly as present in the exhaust of an engine.

sump (US: oil pan) Oil reservoir, as fitted beneath the ***crankcase*** of an engine. See Figure E.1.

sump guard Shield fitted under engine to prevent impact damage to a ***sump*** or ***oil pan***.

sun visor Hinged screen or panel to shield a driver's or passenger's eyes from direct sunlight. Also ***sun shade***.

sun wheel Central spur gear wheel of an ***epicyclic*** gear train. Also ***sun gear***. See also ***annulus***; ***planet wheel***; ***ring gear***. See Figure P.3.

super single Large heavy-duty ***tire*** for commercial vehicle road use, intended to offer the performance of ***double tires*** at lower total weight.

supercharge Artificial increase in pressure of induction air or gases, achieved by pressure charging (or supercharging) from a mechanical air pump. See also ***boost***.

supercharger Mechanical pump or compressor for increasing the pressure of induction air or gases. See also ***blower***; ***turbocharger***.

supplementary driving lamp Lamp to supplement the main or upper beam of a ***headlamp*** in absence of oncoming traffic.

supply dump valve Valve fitted in two-line ***air brake*** systems of tractor-trailer combinations to monitor pressure in respective brake systems and to equalize pressures and ensure parity of braking effort between tractor and trailer.

suppressor Electrical device for preventing radio interference by the radiation of electromagnetic waves from an ignition or other vehicle electrical system.

surface carburetor Early form of carburetor in which the induction air passed over an exposed surface of fuel, collecting as its charge the evaporated volatile components of the fuel.

surface ignition Ignition emanating from ***hot spots*** within an engine cylinder rather than from a timed spark.

surface vaporiser Early form of ***carburetor***, relying on evaporation of gasoline from a large surface. See also ***gauze carburetor***; ***wick carburetor***.

surface-volume ratio Numerical ratio of surface area to volume, specifically in the ***combustion chamber*** of an engine.

surfactant additive Organic fuel ***additive*** that forms a coating on metals or other surfaces, which it protects, acting also as a ***detergent*** or ***dispersant*** by making deposits partially soluble.

surging (1) Fore and aft oscillation of a vehicle. (2) Uncontrolled and transient speed increase in an engine.

suspension Means whereby vehicle body is supported on its undercarriage, comprising springs, dampers and locating linkages.

suspension air control system Pneumatic system for setting the ride height, as of a commercial vehicle trailer ***air suspension***.

suspension geometry The geometric arrangement of the components and linkages of a suspension system and the numerical value of the lengths and angles thereof. See also ***steering geometry***.

suspension rate Change of wheel load at ***center of tire contact*** per unit displacement of the vehicle ***sprung mass***.

suspension roll Roll of the ***sprung mass*** of a vehicle excluding tire deflection.

suzie Coiled, usually colored, plastic tubing that connects the air brake system of a tractor unit with that of its trailer. (Informal) See Figure S.10.

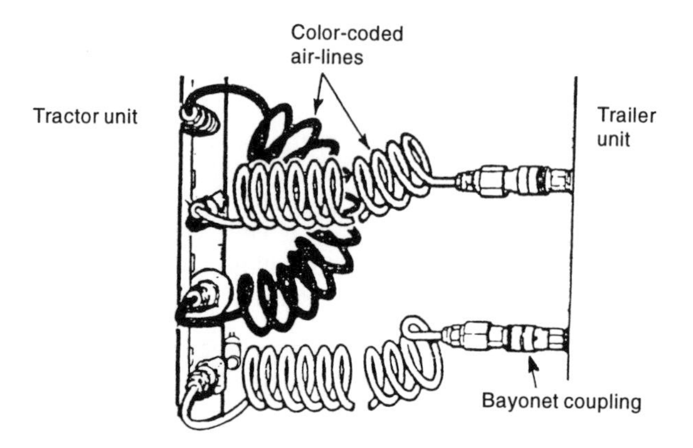

Figure S.10 Trailer brake couplings, or "suzies," of coiled plastic piping, each function identified by color and different size bayonet fitting.

swap body A ***demountable*** freight body of a commercial vehicle or its trailer. Also ***demountable***; ***swop body***. See Figure D.2.

swash plate Rotating disc or plate mounted obliquely as an alternative to the crank in translating reciprocating to rotary motion.

swash plate engine Engine employing a ***swash plate*** rather than a crank to convert piston motion into shaft rotation.

swept circle The diameter of the extreme circle covered by a vehicle turning on maximum lock.

swept volume In a reciprocating engine, the volume of the cylinder formed by the bore diameter and the stroke of the piston. See also ***capacity***; ***displacement***.

swing axle A driven half-axle, pivoted at a central differential case, the change of wheel camber being identical to the angular suspension deflection. A few vehicles have used swing axle front suspension. See Figure S.11.

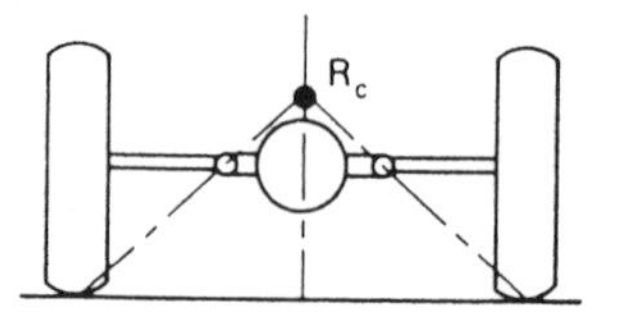

Figure S.11 A rear swing axle showing position of roll center.

swing clearance The diameter of the extreme circle covered by the front corners of a trailer turning with the towing vehicle on maximum lock.

swinging caliper Disc brake ***caliper*** assembly in which the complete assembly can pivot to accommodate wear or distortion, one ***pad*** only being actuated by hydraulic pressure. See Figure S.12.

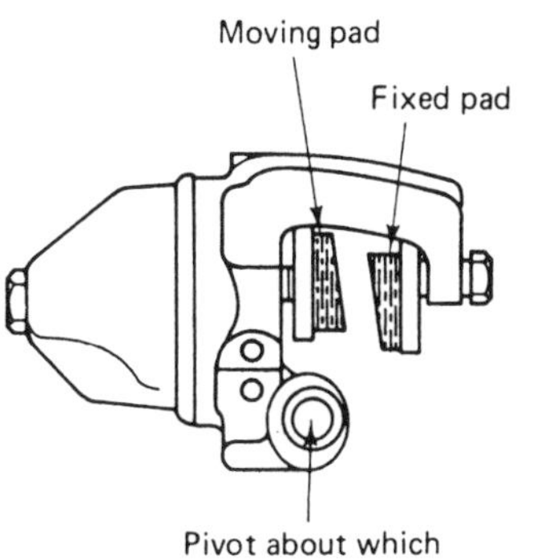

Figure S.12 A disc brake swinging caliper.

swinging shackle Spring shackle pivoted to chassis to support the rear end of a ***leaf spring*** and provide for longitudinal movement of the spring eye. See Figure L.2.

swirl Orderly rotation of combustion gases in an engine cylinder, to improve mixing and heat transfer.

swirl chamber Small chamber or cavity formed in ***cylinder head*** of an engine to promote swirl.

swirl duct Inlet tract of an engine formed to promote ***swirl***, particularly where the main flow of ***ram air*** is provided by a separate duct. See also ***ram duct***.

switch terminal Flat electrical terminal, in line with run of cable. For heavier currents usually with hole for positive attachment by bolt or screw.

swivel pin A ***kingpin***.

symmetrical beam Light beam in which both sides of the beam are symmetrical with respect to the median vertical plane of the beam.

synchromesh Change-speed gear system in which the speed of rotation of a selected gear is automatically synchronized with that of the mainshaft immediately prior to engagement. See Figures G.1 and S.13.

synchromesh gearbox Change-speed transmission in which the engagement of gears is mechanically synchronized, thus obviating the need for ***double clutching*** (declutching). See Figure G.1.

synchromesh sliding sleeve Internally splined sleeve that performs the function of a ***dog clutch*** by mechanically locking the mainshaft gear pinion to the ***balk ring*** and mainshaft when gear is engaged in synchromesh gearbox. See Figure S.13.

synchronizer Sliding clutch mechanism by which gear engagement is synchronized in a synchromesh gearbox. See Figure S.13.

synlube A synthetic lubricant. (Informal)

synthetic lubricant Lubricant oil using chemically derived rather than natural oil base stock.

synthetic rubber An artificially produced resilient material with properties similar to natural rubber. There are many types, such as ***butyl rubber***, ***neoprene rubber***, ***nitrile rubber***. See also ***elastomer***.

system protection valve ***Air brake*** safety device that permits other reservoirs to be partly charged in the event of failure of one reservoir.

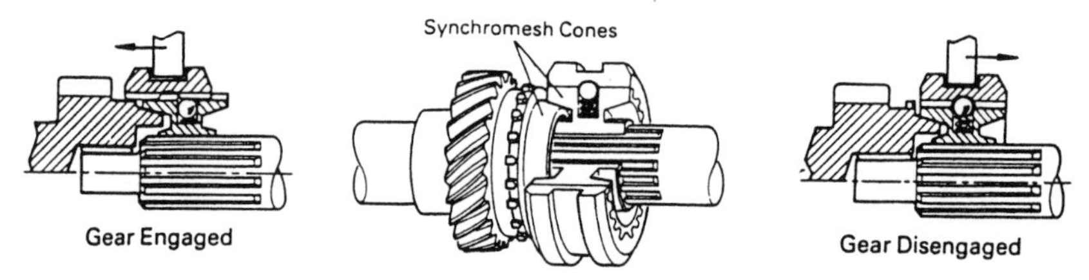

Figure S.13 Synchronizers.

T

T head engine Side-valve engine with ***inlet valves*** and ***exhaust valves*** on opposite sides of the cylinder block, giving a ***combustion chamber*** resembling the shape of the letter T. See Figure S.5.

T slot piston Piston with vertical slots in the ***skirt*** terminating in horizontal slots to reduce heat transfer and counteract thermal distortion. See also ***split skirt piston***.

tachograph Instrument that records vehicle usage data against a time base, usually by stylus on a paper disc.

tachometer Instrument for measuring speed of rotation, as of an engine.

tag axle A non-driven axle of a ***tandem axle*** suspension of a commercial vehicle, following the driven axle. See also ***pusher axle***.

tag tandem Commercial vehicle tandem axle of which only the forward axle is powered, the after axle being a trailing or ***tag axle***.

tail fin See ***fin***.

tail lamp Red lamp to show the presence of a vehicle from the rear. May incorporate brake warning lamp.

tail lift Lifting platform fitted to the rear of a goods vehicle to facilitate loading and unloading.

tail pipe The rearmost pipe of an ***exhaust system***, downstream of the rear ***muffler (silencer)***. Occasionally used to describe intermediate pipes. Also ***kick-up pipe*** (US informal); ***tail pipe***. See Figure E.2.

tail pipe emissions Emissions from an engine's tail pipe, as opposed to evaporative and other emissions.

tailgate Hinged transverse rear door of a van or passenger car. See also ***hatchback***.

tailshaft An output or drive shaft from engine, gearbox or other item of rotating machinery.

tailshaft governor Speed-sensitive device that monitors ***torque convertor*** tailshaft speed and governs engine speed accordingly.

take-off See ***power take-off***.

TAME See ***tertiary amyl methyl ether***.

tandem axle Undercarriage arrangement where two or more axles are close coupled, for example the paired ***steered axles*** or rear axles of a heavy commercial vehicle. See also ***dual-drive tandem***; ***interactive suspension***; ***pusher tandem***; ***tri-axle***. See Figure T.1.

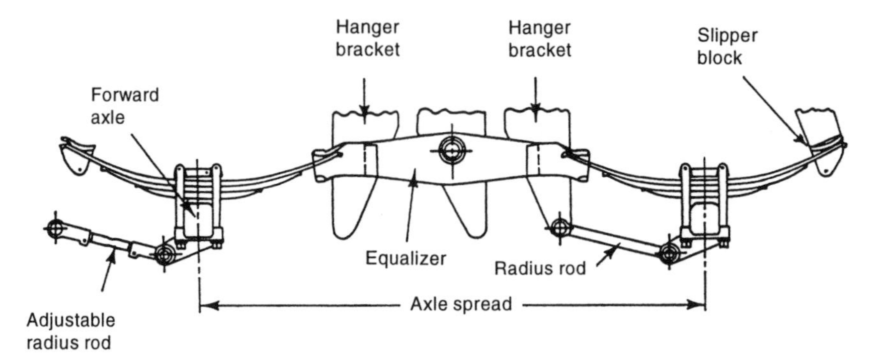

Figure T.1 An equalizing tandem-axle leaf spring suspension with equalizer or equalizing beam (Crane Fruehauf).

tandem master cylinder Two master cylinders in one housing for operating ***divided system brakes***. See Figure T.2.

tangential force variation See ***tractive force variation***.

tank (1) A closed container or reservoir from which a liquid may be drawn, as gas tank (US) or petrol tank (UK). (2) Track-laying armored fighting vehicle.

tank body See ***tanker***.

tanker Commercial vehicle designed for the transport of bulk liquids or powders. See Figure A.4.

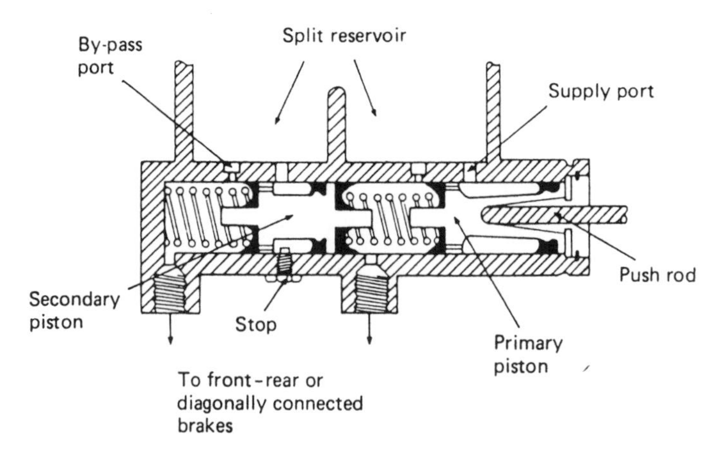

Figure T.2 A brake tandem master cylinder.

taper leaf spring Single leaf spring of tapering section, and usually of high capacity for weight. Also tapered leaf spring. See Figure A.6.

taper roller bearing Rolling element bearing employing conical rollers and able to carry thrust (axial) loads as well as journal loads. See Figure D.5.

tappet (1) **(US: valve lifter)** Cylindrical reciprocating ***cam follower***, as for example in an engine ***gear train***. The tappet converts cam ***lift*** into linear reciprocating movement which it transmits directly or indirectly to the valve. Valve plunger. (Obsolescent) (2) The sliding component that separates the brake ***shoes*** in a ***wedge brake***. (3) Adjusting screw for valve clearance. (US) (4) A valve ***rocker arm***.

tare weight Weight of an unladen commercial vehicle without fuel, water and oil, but subject to local definition. See also ***dry weight***, ***kerb weight***.

targa top Removable rigid roof-panel, particularly of a ***sports car***.

tarp hook Hook or cleat for fastening a tarpaulin or fabric cover over an open truck body, or for fastening the sides of a ***curtain-sider***. (Informal)

Tautliner® A type of ***curtain-sider*** commercial vehicle employing a proprietary tensioning system and other features.

Taxation Weight Weight, particularly of a goods vehicle, as defined for taxation purposes.

taxi Passenger car, often of special design, licensed to carry passengers for hire. Also ***cab***; ***hackney*** (UK); ***taxicab***.

taximeter Clock or electronic display showing the fare incurred on a ***taxi*** journey.

TC Twin carburetor. Sometimes ***twin camshaft***.

TCP See ***Tri-cresyl phosphate***.

TEL See ***Tetra-ethyl lead***.

telescopic damper See ***telescopic shock absorber***.

telescopic fork Steering/suspension arrangement mainly applied to ***motorcycles***, in which the steered wheel is straddled by two telescopic struts linked below the ***steering head***. See also ***girder fork***.

telescopic shock absorber Coaxial tubular hydraulic damper. See Figure S.4.

Temperature Grade Letter code on a ***tire*** indicating resistance to degradation due to heat build-up.

temporary switch-off Facility for switching off an engine, followed by spontaneous restarting to reduce emissions, for example when waiting in traffic.

tension rod Steering ***track rod*** or tie rod.

terminal post Tapered barrel or lug-shaped terminal on ***battery*** to which battery lead is attached by clamp or screwed cap.

tertiary amyl methyl ether Oxygenate used as a gasoline blend component. Also ***TAME***.

test cell A room equipped for testing, as of an engine or other components. See also ***production test cell***.

test cycle Laboratory or highway test procedure that follows a strictly controlled sequence of operating parameters, usually to simulate driving under realistic conditions and to assess performance in respect of ***emissions***, fuel consumption or other measurable quantities.

test drum Rotating cylindrical flywheel of specified size, as for example used in the testing of ***tires***.

Tetra-ethyl lead. An ***anti-knock*** fuel ***additive***. Also ***TEL***.

Tetra-methyl lead. An ***anti-knock*** fuel ***additive***. Also ***TML***.

THC Total ***hydrocarbon*** (content of exhaust).

thermal efficiency Ratio of useful work performed by an engine to the total energy content of the fuel consumed. A measure of the efficiency of the combustion process.

thermal loading Total effect of heat and temperature on mechanical and structural components, particularly of an engine. Not a load in the quantitative mechanical sense.

thermal reactor Emission control device relying principally on high temperature to effect oxidation of pollutants.

thermodynamic cycle Idealised sequence of operating stages of engine or other heat machine, often illustrated by a diagram showing pressure plotted against displacement volume. Actual engine cycles of operation differ considerably from their ideals. Representative cycles are the ***Diesel cycle***; ***Otto cycle***; ***Rankine cycle***.

thermoforming Means of production of formed panels from plastics sheet, such as ABS, by heat-softening flat sheets of material, prior to forming by a process such as ***vacuum forming***.

thermostatic interrupter Electrical device to prevent overheating in ***lighting*** circuits in event of a fault or short-circuit.

thermosyphon ***Radiator*** cooling system that relies on the density differential between hot and cooled water to produce circulation of coolant.

thin wall bearing A journal ***shell bearing***, of thin section relative to width and diameter.

third differential Differential between two ***driving axles*** of a multi-axle vehicle, to accommodate differences in axle speed between the two axles, often with facilities for locking to ensure equal driving speed.

thixotropic Of a material such as a paint, solid in undisturbed state, but fluid when stirred.

thrash Severe transverse oscillation or whip of a chain or belt drive system. See also ***damping slipper***.

three axle rigid Commercial vehicle with load-carrying ability and three axles, one, two or all of which may be driven. See also ***Chinese six***; ***rigid truck***. See Figure C.9.

three bearing crankshaft Engine crankshaft having one additional central journal bearing. Common feature of lightly loaded four-cylinder engines.

three brush dynamo Obsolete type of DC generator in which a third brush controls output current.

three-piece construction Construction of an item in three usually separable pieces, but particularly a heavy vehicle ***wheel*** in which a removable ***ring flange*** and locking ring locate and restrain the ***tire bead***.

three-quarter floating axle Axle arrangement in which the ***axle*** or ***half shaft*** reacts cornering loads but carries only a small proportion or none of the vehicle's weight, so that, in the event of half shaft failure the wheel bearing would still retain the wheel.

three-way catalyst See ***three-way converter***.

three-way converter Catalytic converter containing one stage coated with platinum and palladium and another with platinum and rhodium for the control of ***hydrocarbons***, ***carbon monoxide*** and ***oxides of nitrogen***. Also ***dual-bed converter***; ***TWC***. See also ***two-way converter***.

three-way tipper Tipper body so designed as to allow the load to be discharged to either side, or to the rear.

three wheeler (US: tricar) Three-wheeled motor vehicle other than a motorcycle combination.

throat Narrow end of a tapered aperture or venturi, as of a ***carburetor*** inlet.

throttle (1) Valve, particularly a butterfly valve, for controlling the flow of fuel through a carburetor. See Figure C.3. (2) A venturi or stepped pipe for throttling a supply of gas or vapor. (3) The ***accelerator*** pedal, being the control that operates the throttle. (Informal)

throttle body injection Fuel injection into a venturi device similar to a carburetor barrel. See also ***downstream injection***; ***multi-point injection***.

throttle body injection system Electronically controlled closed-loop single point gasoline fuel injection system.

throttle cable Cable by which a throttle is operated, as from the twist-grip of a motorcycle.

throttle lever Lever for operating a ***carburetor*** throttle.

throttle stop Regulator, usually an adjusting screw on a ***carburetor***, that limits the closure of the throttle, and thus sets the ***tick-over*** speed of an engine.

throttle valve Valve by which the flow of air into a ***carburetor*** is regulated, and thus the output of the engine. See also ***butterfly valve***. See Figure C.3.

throttling Reducing the power output of an engine by closing the throttle.

throw (1) Connecting rod or big-end bearing of a ***crankshaft***. (Archaic, except US) (2) Radial distance between the ***crankshaft*** and ***connecting rod*** bearing axes, equivalent to half the ***stroke***.

throwout fork (UK: clutch release lever) Lever, mechanically or hydraulically actuated by ***clutch*** pedal, that acts on the throwout (clutch release) bearing to disengage clutch. Also ***thrust bearing actuating lever***. See also ***pull type clutch***; ***push type clutch***. See Figure C.5.

throwout sleeve Sliding sleeve on which ***clutch release bearing*** (throwout bearing) is mounted.

thrust bearing Bearing intended principally to react axial or thrust load. Also ***thrust race***. (Informal)

thrust bearing actuating lever Lever, often fork-ended, by which the ***clutch*** is disengaged. A ***clutch release lever*** or ***throwout fork***. (US informal) See Figure C.5.

thrust race See ***thrust bearing***.

thump Periodic vibration or audible sound emanating from ***tire***, producing a pounding synchronous with wheel rotation.

tick over Running of an engine at lowest practical speed with drive disengaged. Also ***idle speed***. See also ***idling***.

tickler Plunger, usually sprung, by which the ***float*** of a ***carburetor*** can be manually depressed to prime the carburetor prior to starting the engine.

tie bar Any bar that serves to tie parts or components together. See also ***strut***.

tie rod (1) Any structural member or mechanical linkage that is normally in tension. (2) Steering track rod or cross rod, or any nominally transverse linkage that directly or indirectly actuates ***steered wheels***. See Figure S.8.

tilt-bed trailer Trailer with facilities for tilting of the platform to allow loading and unloading from a ramp. See also ***beaver tail***.

tilt cab Commercial vehicle ***forward control*** cab hinged so that it can tilt forward to facilitate access to engine and running gear.

tilt deck Commercial vehicle, ***semi-trailer*** or ***full trailer*** of which the whole or part of the ***cargo*** floor or ***platform*** may be tilted relative to the ***chassis*** for loading and unloading.

timing Scheduling of events related to an engine's operating ***cycle***, such as the opening and closing of ***valves*** (valve timing), or the firing of the mixture by spark (***ignition timing***).

timing belt Flat-toothed belt of reinforced rubber for driving the ***camshaft*** of an engine from the ***crankshaft***. See Figure E.1.

timing chain Continuous roller chain for driving the ***camshaft*** of an engine from the ***crankshaft***.

timing gear (1) Gear wheel (or wheels) for driving the ***camshaft*** of an engine from the ***crankshaft***. (2) Any mechanism associated with the driving of a timing mechanism.

timing mark Any mark on ***flywheel***, ***crankshaft***, ***camshaft*** or other rotating component to serve as reference when the valves, ignition or injection system of an engine are being timed. Also ***reference mark***.

timing scatter Deviation of the actual from the intended or optimum spark event. A characteristic problem with mechanical/pneumatic distributors at higher speeds.

timing shaft (UK: distributor shaft) Shaft for driving an ***ignition distributor***. See also ***quill drive***.

Timken Test A test which measures the wear prevention and EP characteristics of ***oils*** and ***greases***.

tipper (US: dump truck) Rigid commercial vehicle or trailer with tilting body or hopper and opening taildoor through which bulk materials may be discharged. See also ***three-way tipper***. See Figure R.6.

tipping body Body type of a vehicle intended for tipping or discharge of bulk materials.

tipping gear Machinery, usually hydraulic, for operating a tipping body of a ***tipper***. Term sometimes implies the tipping body in addition to its machinery and controls.

tire (UK: tyre) (1) Air-filled or solid covering for a wheel, normally of rubber. (2) Device made of rubber, fabric and other materials which, when filled with fluid under pressure and mounted on a wheel, cushions and sustains the load. See also ***inner tube***. See Figures B.3 and R.2.

tire carcass See ***carcass***.

tire casing (1) The tread and shoulder of a tire. (2) The rubber-bonded cord structure of a tire. See Figures B.3 and R.2.

tire contact center See ***center of tire contact***.

tire face Side of tire facing away from the vehicle, if fitted as recommended.

tire load Load or weight supported by a tire.

tire load rating Maximum recommended load for a given tire at specified cold inflation pressure.

tire power loss Mechanical power input converted into heat by the tire.

tire pull force Tendency of a tire to pull to the left or right when running on a flat, even surface. This tendency is not related to any characteristics of ***steering*** or ***suspension geometry***. See also ***conicity***; ***plysteer***.

tire radial run-out Variation in extreme radius of the ***tread*** of a mounted tire in the ***wheel plane***.

tire scrub Sliding movement of tire at right angles to the plane of the wheel, effectively the active component of ***slip angle***.

TML See ***Tetra-methyl lead***.

toe-in (1) Setting of paired wheels on an axle so that the leading edge of each wheel is inclined slightly inward. Also ***gather***. (2) The numerical value of such inclination, usually expressed as the difference in effective track from leading to trailing edge of wheel at specified diameter, usually at the ***rim flange***.

toe-out Setting of paired wheels on an axle so that the leading edge of each wheel is inclined slightly outward. See also ***toe-in***.

toluene Aromatic compound used as a ***solvent***, and, because of its octane qualities, as a gasoline blend component.

Toluene Number An early ***anti-knock*** index of fuel rating, superseded by ***octane number***.

tonality (1) The pattern of frequencies that make up a sound. (2) Quantifiable structure of sound, such that a pure fundamental had 100 percent tonality and white noise zero tonality. (3) Noise, as for example of a tire, associated with fundamental frequency and its harmonics.

tonneau Passenger compartment of an early car, particularly when separated from the driver's compartment.

tonneau cover Detachable fabric cover to protect the passenger compartment of an open car when not in use. Also ***tonneau***. (Informal)

tonneau panel External rear side panel, particularly of a four-door ***saloon*** or ***sedan*** car. (Mainly UK terminology) See Figure B.4.

toothed belt Positive action reinforced rubber belt in which parallel striations engage with grooves in a driving and driven wheel. Cogged belt. (Informal) Commonly used for valve ***timing gear*** as an alternative to a roller chain, when it is often called a ***timing belt***. See Figure E.1.

top dead center Uppermost point of movement of ***piston*** in a ***cylinder***, or the point furthest from the crankshaft axis. Also ***TDC***; ***upper dead center***; ***outer dead center***.

top sleeper A sleeper cab arrangement of a commercial vehicle, where sleeping accommodation for the driver is provided in a pod above the ***cab***.

top speed Maximum steady speed of a vehicle, usually measured on a flat surface and with no wind.

top yoke Structural item that links the tops of the ***telescopic forks*** of a ***motorcycle*** front suspension and pivots on the uppermost bearing of the ***steering head***. See also ***bottom yoke***.

toric transmission Generic term for constantly variable transmissions that operate by changing the input and output radii of discs or wheels rolling within a split torus. See also ***Hayes transmission***; ***Perbury drive***.

torque arm (1) Any arm or lever primarily intended to resist or transmit torque. (2) Suspension linkage that reacts drive and braking wind-up of an axle.

torque ball Spherical housing for ***universal joint*** in some designs of ***torque tube*** transmission.

torque converter (1) Hydraulic torque converter, consisting of rotating and static vane assemblies, by which torque can be transmitted, multiplied and controlled. A feature of many ***automatic transmissions***. See also ***fluid flywheel***; ***lock-up***. (2) Device that transmits torque from one shaft to another usually by hydraulic means and permitting asynchronism between the shafts. See Figure T.3.

torque converter pump Vaned driven rotating element of a torque converter, normally set opposite to the turbine. See also ***stator***. See Figure T.3.

torque fluid A low-viscosity, high-stability oil used in ***torque converters***.

torque ratio Ratio of output torque to input torque, as of a torque converter. Often expressed as a percentage.

torque tube Tube enclosing ***propeller shaft*** and forming torsionally rigid connection between gearbox and final drive. See Figure T.4.

torque tube transmission Vehicle final drive system in which the ***propeller shaft*** (drive shaft) is housed within a structural tube which is rigidly attached to the ***differential*** housing. The ***back axle*** housing, differential case and torque tube therefore form one rigid assembly, thus eliminating one of the ***Hotchkiss drive's*** universal joints and divesting the rear suspension of the need to react drive and braking torques and forces. See Figure T.4.

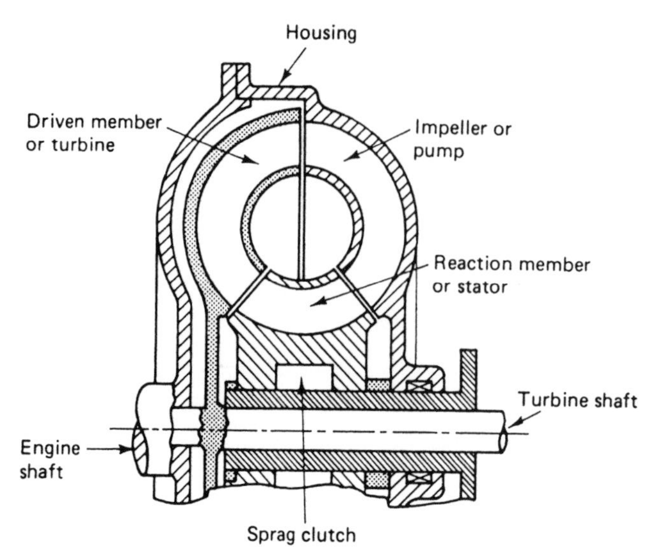

Figure T.3 Components of a torque converter.

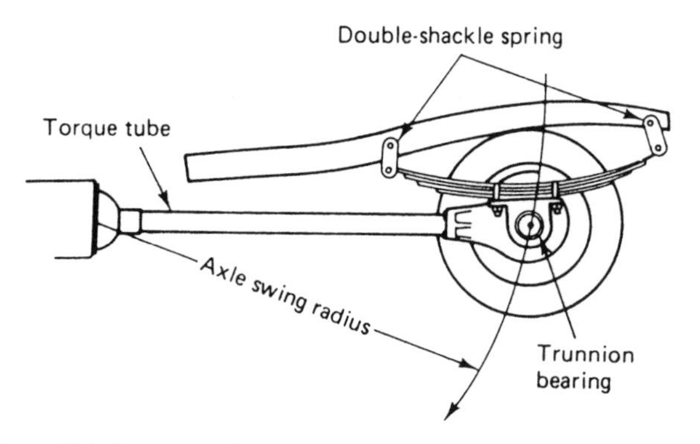

Figure T.4 A torque tube transmission with semi-elliptical leaf springing.

torsen differential Helical/skew gear differential providing limited slip characteristics. Contraction of torque sensing.

torsion bar Bar or rod that fulfills a mechanical or structural function by reacting loads in torsion. See Figure T.8.

torsion bar suspension Suspension that uses a rod, bar or tube in torsion as a spring.

torsional damper See ***torsional vibration damper***.

torsional vibration damper (1) Any mechanical or hydraulic device that reduces torsional vibrations, as of an engine ***crankshaft*** or ***transmission shaft***. (2) Lever-type suspension damper or ***shock absorber***.

torso line Line on a two-dimensional drafting template connecting the shoulder reference point with the ***H-point***. See ***waistline***.

torus (1) A ring, usually of thick, circular section. (2) A flexible coupling of toroidal shape.

total driving resistance See ***driving resistance***.

tow bar Rigid bar by which a trailer or towable vehicle is towed. A ***drawbar***.

tower wagon Vehicle equipped with a usually retractable tower mechanism to provide access to overhead cables, street lamps, etc.

towing fork (US: trailer hitch) Mechanical attachment whereby a trailer is coupled to a tractor. Also ***towing jaw***.

towing hook Vertically oriented hook, attached to the chassis or rigid structural member of a towing vehicle, with which the towing eye of the towed vehicle engages. See Figure T.5.

track (US: tread) Transverse distance between left and right side wheels on the same axle, measured between specified points, such as the ***centers of tire contact***, or the centers of the widths of the tracks of a ***track-laying vehicle***. See also ***track gage*** (track gauge). See Figure T.6.

track arm Pivoted transverse member constraining the wheel against transverse motion, particularly in a ***MacPherson strut*** or ***Chapman strut*** independent sus-

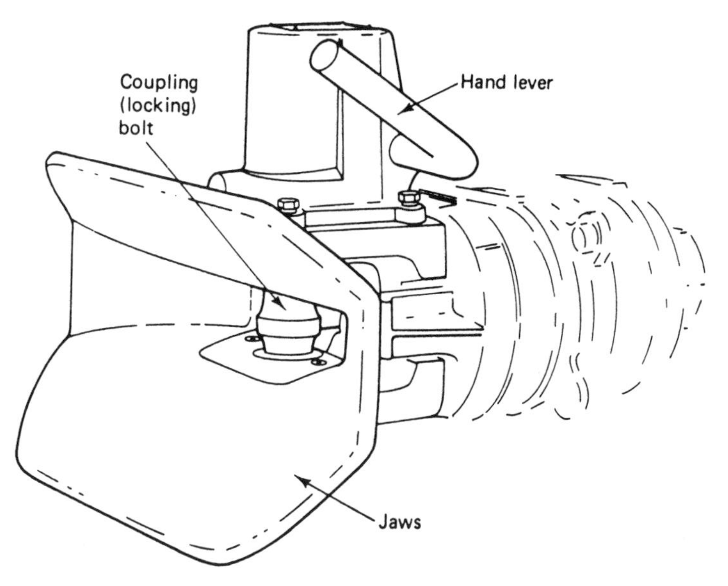

Figure T.5 A trailer coupling jaw or towing hook, showing the coupling or locking bolt with which the drawbar eye engages.

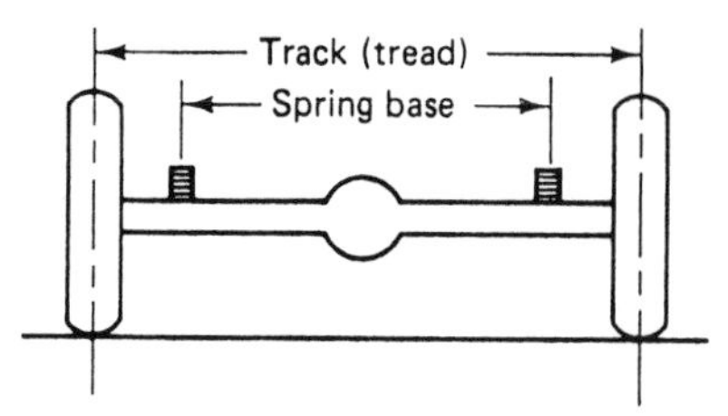

Figure T.6 Wheel track (tread) and spring base.

pension system. Also track control arm. See also ***tie rod***. Term is sometimes, though misleadingly, used as an alternative to track rod. See Figure M.1.

track chain Continuous linked track of a tracked or ***track-laying vehicle***, such as a tank or crawler tractor. See also ***sprocket***; ***idler***. See Figure C.12.

track control arm Any transverse member that maintains the ***track*** or reacts lateral loads in a vehicle suspension. A ***tie rod***.

track gage (UK: track gauge) Lateral distance between mid-chord points of the tracks of a ***track-laying vehicle***.

track-laying vehicle Vehicle of which the wheels run within a continuous chain or track, as for example a fighting tank. A ***half-track vehicle*** is also usually considered to be a track-laying vehicle. A ***crawler***. (US informal) See Figure C.12.

track pin Metal pin that forms the pivot and coupling between two track elements of a ***track chain***. See Figure C.12.

track pitch Distance between track pin centers of a ***track chain***. See Figure C.12.

track rod Transverse rod that connects the ***steering arms*** of ***steered wheels*** and so maintains the geometric relationship between them when steering. A ***rack and pinion*** steering system will have two ***track rods*** or side rods. In an ***independent front suspension*** steered from a ***steering box*** there will often be a single ***intermediate rod*** or ***relay rod*** (US) communicating the steering to the ***steering arms*** by two side rods or tie rods. See also ***relay rod***; ***tie rod***. See Figure S.8.

track rod end Ball joint forming pivot between ***track rod***, or ***tie rod***, and ***steering arm***.

track rod lever A ***steering arm*** or ***Pitman arm***, or other lever actuating a track rod. See Figure S.8.

track rollers The bearing and guiding wheels that support the ***track*** of a ***track-laying vehicle***. See Figure C.12.

track-shoe Load-bearing link of the track chain of a ***track-laying vehicle***. See also ***grouser***. See Figure C.12.

track thrash Periodic whipping motion of the track of a ***track-laying vehicle***.

tracked vehicle See ***track-laying vehicle***.

tracking (1) Conducting of high-voltage electricity along the surface of an insulator such as a cable or inside of ***distributor cap***, due particularly to accumulated dirt and dampness. (2) The geometric settings and alignments of a vehicle ***steering*** and ***suspension*** system, or the adjustment thereof.

tracking error Misalignment of road wheels particularly where adversely affecting ***steering*** or tire wear.

Tracta joint See ***Bendix-Tracta joint***.

traction Driving force or effort of a motor vehicle, or its pulling power. See also ***lugging***.

traction avant Front-wheel drive. (French)

traction bar Rod or strut attached to a ***live axle*** to react torque loads and prevent or minimise ***axle wind-up*** and ***hop***.

traction control System that prevents the spinning of driven wheels of a vehicle when excess power is applied.

traction differential See ***limited slip differential***.

traction engine Heavy road ***locomotive*** usually steam powered, formerly used for heavy haulage, and the hauling and operation of farm machinery.

Traction Grade Letter code of a ***tire*** representing the ability to stop a vehicle on a wet surface.

tractive effort Quantitative measure of the ability of a vehicle to pull or haul.

tractive force variation Variation of driving or tractive force exerted by a tire per revolution. Also ***tangential force variation***; ***TFV***.

tractive unit The non-payload-carrying towing vehicle of an articulated commercial vehicle. Also ***tractor*** or ***tractor unit***. See Figure A.4.

tractor (1) A steerable vehicle for towing, but particularly an agricultural tractor. (2) A commercial vehicle equipped with ***fifth-wheel*** for towing a ***semi-trailer***. A ***semi-tractor***. (Mainly US usage) Also ***tractor unit***. (Informal)

traffic indicator (US: turn signal lamp) Lamp to show the intention of a driver to turn. Also ***direction indicator lamp***; ***flasher***. (Informal)

trafficator Mechanically or electrically raised arm, sometimes illuminated, and mounted at the side of a vehicle to indicate the driver's intention to turn. Also ***semaphore indicator***. (Obsolete)

trail Horizontal distance between a vertical line through the front wheel centerline and the projection of the ***steering head*** or ***kingpin axis*** measured at ground level.

trailer A non-powered vehicle designed to be towed by a tractor. See also ***drawbar trailer***; ***full trailer***; ***semi-trailer***.

trailer hitch (UK: towing fork or towing jaws) Mechanical link by which a trailer is towed by a towing vehicle. See Figure T.5.

trailer sway (UK: trailer swing) Excessive directional deviation of a trailer, particularly a ***semi-trailer***, from the intended direction of tow resulting from ***slip*** of the trailer rear wheels, either as movement in one direction or as an oscillation. See Figure T.7.

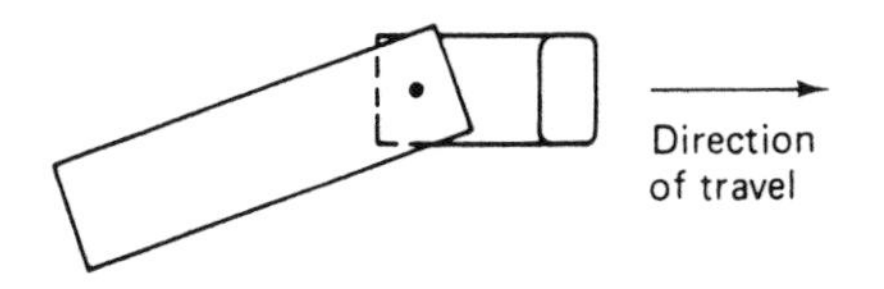

Figure T.7 Trailer sway (trailer swing) of an articulated vehicle.

trailing arm Suspension linkage supporting wheel assembly aft of a transverse ***pivot axis***. See Figure T.8.

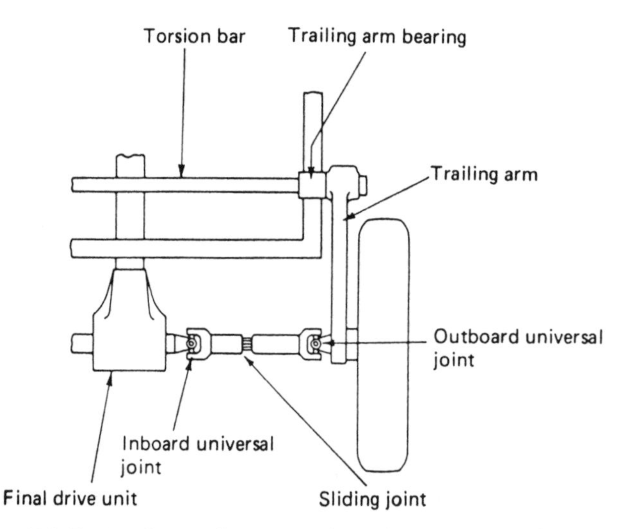

Figure T.8 Torsion bar trailing arm independent rear suspension.

trailing axle The non-driven or dead after axle of a ***tandem axle*** undercarriage.

trailing link A ***trailing arm***.

trailing shoe (US: secondary shoe) Shoe of a ***drum brake*** system in which the actuated end trails, and the leading edge faces the normal direction of rotation. See Figure C.1.

tram See ***tramcar***.

tramcar (US: streetcar) Passenger-carrying rail vehicle, usually electrically propelled, that runs on tracks laid within a street system. Also ***tram***. (UK informal)

tramp ***Wheel hop*** in which a pair of wheels hop in opposite phase.

tramway Track on which a ***tram*** runs.

transaxle (1) Rear axle assembly that incorporates the main ***change-speed gearbox***. (US) (2) Combined gearbox and ***differential*** unit attached to the engine in some ***front-wheel drive*** vehicles. Contraction of transmission-axle. (3) Transmission which provides differential output.

transfer box (US: transfer case) Gear system that apportions the drive between front and rear axles of a ***four-wheel drive*** system, thus having two output shafts, and one input shaft. Also ***transfer gearbox***. (UK) See also ***full-time drive***; ***part-time case***. See Figure T.9.

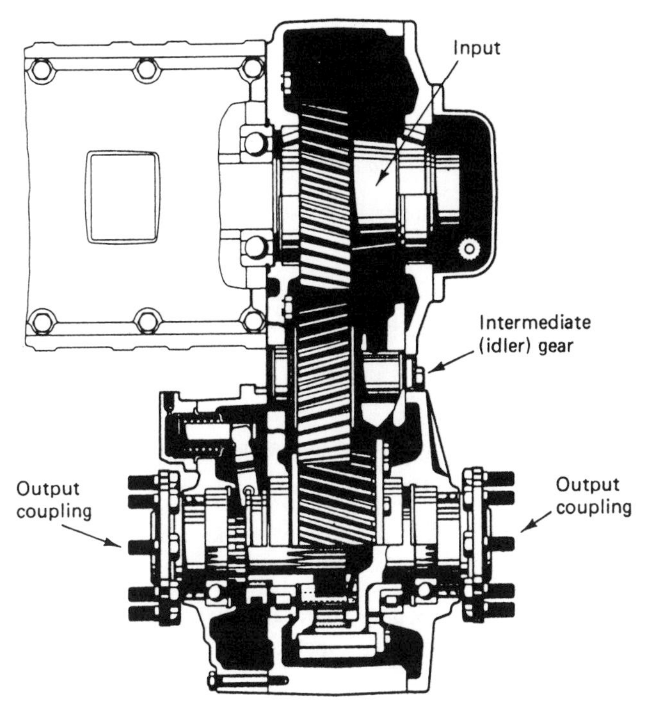

Figure T.9 A transfer case or drop box (ZF).

transfer drive Bevel gearbox, the output axis of which is at an acute angle to the input axis, mainly used as a drive from the gearbox to the ***propeller shaft*** in transverse rear-engined buses.

transfer gear (1) Transmission that apportions an input drive between two shafts, as for example between front and back axle in ***four-wheel drive***. (2) A ***drop gear***. (Mainly US usage)

transfer port Aperture in cylinder of ***two-stroke engine*** through which compressed combustible mixture of fuel and air passes under pressure from ***crankcase***. See Figure T.11.

transistorized ignition (1) Conventional mechanical ***contact breaker*** ignition system with transistor voltage regulation or control permitting higher speeds. (2) ***Breakerless ignition*** system in which a transistorized electronic circuit triggers the spark. See also ***distributorless ignition system***; ***electronic ignition***; ***solid-state ignition***; ***triggering***.

transition system Metering or flow system in a ***carburetor*** that provides smooth engine operation under transient conditions and throttle openings, as for example ***acceleration*** or ***overrun***.

transmission (1) **(UK: gearbox)** Mechanical unit containing a manual or automatic change-speed gear system and associated actuating machinery. See Figure G.1. (2) Collective term for the components, such as ***clutch***, ***gearbox***, ***drive shaft***, whereby power is transmitted from engine to driven wheels. (Mainly UK usage) See also ***drive line***; ***powertrain***.

transmission brake Brake operating on the transmission system of a vehicle rather than directly on the wheels. See also ***inboard brake***.

transmission loss That part of the engine brake horsepower that is absorbed by the ***clutch***, ***gearbox*** and other transmission items.

transmission shaft (UK: propeller shaft) Shaft that transmits the rotary power from an ***engine*** or ***gearbox*** to a driven axle. See Figure P.5.

transmission tunnel Raised section along centerline of vehicle ***floor panel*** to accommodate the ***transmission shaft***. See Figure B.4.

transporter Usually articulated commercial vehicle built for the transport of other vehicles, or of abnormally large loads.

transverse engine Engine having ***crankshaft*** axis athwart the vehicle, at right angles to the direction of motion.

transverse link Any mechanical linkage, particularly in a ***suspension system***, that is located in a transverse plane, and provides constraint in transverse or trackwise movement.

transverse springs Springs, but particularly ***semi-elliptical leaf springs***, located in a transverse plane.

trapped-line pressure valve See ***check valve***.

travel trailer Box-trailer for towing behind a passenger car.

tread (1) The portion of a ***tire*** that comes into contact with the road. (2) The grooved or patterned face of a tire, intended for road contact. See also ***tread depth***. See Figures B.3 and R.2. (3) **(UK: track)** Transverse distance between left and right side wheels on the same axle, measured between specified points, such as the ***centers of tire contact***. See Figure T.6. (4) **(UK: track)** Transverse distance between the centers of the tracks of a ***track-laying vehicle*** or crawler.

tread arc width Peripheral distance between the two ***shoulders*** of a ***tire***, measured along the tread contour. For tires with round shoulders the distance is measured between the points of contact of the imaginary extensions of tread and ***sidewall*** profiles.

tread contact length Length of a tire ***contact patch*** or ***footprint***, usually measured in the wheel plane. See also ***tread contact width***.

tread contact width Distance between the extreme edges of tire-road contact at a specified load and pressure measured parallel to the wheel axis at zero slip angle. See also ***tread arc width***.

tread contour Cross-sectional shape of the tread surface of an inflated unloaded tire neglecting tread pattern depressions.

tread depth Distance between the base of a tire tread ***groove*** and a line tangential to (or in simple measurements joining) the adjacent surface of the tire.

tread noise Noise produced by the interaction of tire tread pattern and the road surface, though excepting ***squeal*** and ***slap***.

tread pattern The distinctive geometric pattern of a tire's ***tread***.

tread profile The general form of the tread of a ***tire***, viewed in section. See also ***tread radius***. See Figures B.3 and R.2.

tread radius Radius of a tire's tread measured in the plane of the ***spin axis***.

tread separation Separation of tire tread from the body of the tire.

tread shoulder Extreme edge of a tire tread, particularly when coming to a distinct rim. See also ***shoulder rib***. See Figures B.3 and R.2.

Treadwear Grade Numerical code representing wear resistance of a tire.

trembler coil Electrical coil for generating high tension electricity from a low voltage source. (Obsolete)

trencher (1) Mechanical tool for continuous cutting of a trench, often equipped with means of displacing the excavated material. (2) Wheeled or track-laying vehicle equipped with a trenching tool.

tri-axle Axle arrangement, particularly on a ***semi-trailer***, in which three axles are located in close succession as a means of distributing weight. See also ***tandem axle***. See Figure T.10.

Tri-cresyl phosphate Formerly used as a fuel ***additive*** to minimise misfire and ***pre-ignition*** in leaded gasolines. Also reduces valve wear in ***unleaded*** gasolines. Also ***TCP***.

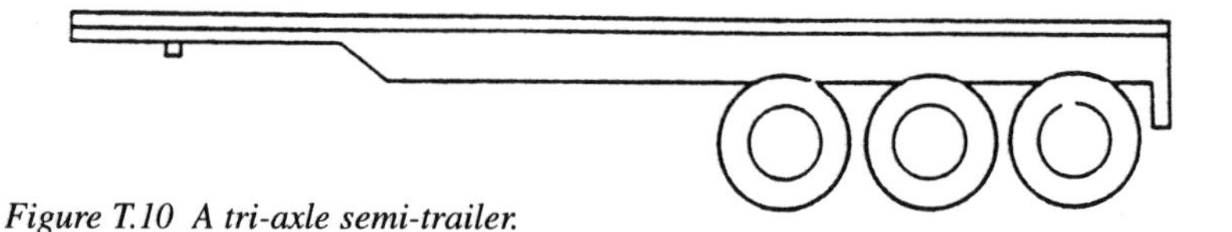
Figure T.10 A tri-axle semi-trailer.

tri-pot joint A radially self-supporting universal ***plunging joint*** having three rollers on radial pins. See also ***pot joint***. See Figure P.4.

tricar (UK: three wheeler) A three-wheeled passenger car or van.

trigger Any device, but particularly an electrical, optical or electronic device, that initiates an event in an electrical or electronic circuit.

trigger rod Hand-lever-operated rod that disengages the locking pawl on a ***handbrake***.

triggering Initiation of a spark in an ***ignition system***, as by the action of a contact breaker. See also ***electronic triggering***.

trip computer Dedicated microcomputer circuit that monitors various performance parameters to give readings of, for example, average speed, fuel consumption, fuel cost per mile.

trip recorder Secondary mileometer that may be manually set to zero to record journey length. Also ***trip odometer***.

triple-reduction axle Heavy vehicle axle with three stages of reduction gearing between ***propeller shaft (transmission shaft)*** and ***final drive***.

Triplex A high-strength safety glass (trade name), originally of laminated construction.

trochoidal rotor Working rotor of the ***Wankel engine***, of triangular form with convex sides.

trolley bus Electrically powered bus equipped to draw current from overhead wires while in motion. Also ***trolleybus***.

trough method A range/mean method of fatigue analysis.

truck (UK: lorry) A commercial goods vehicle, other than a light delivery vehicle.

truck cab The enclosed driver's compartment of a commercial vehicle.

truck chassis Incomplete but mobile truck unit including ***frame***, ***axles***, ***suspension***, ***engine*** and ***transmission*** and related mechanical components, but excluding ***cab***, ***wings*** and ***bodywork***. See also ***chassis cab***.

truck-tractor Heavy vehicle equipped for towing a ***semi-trailer*** by supporting its forward end on a ***fifth-wheel***. See also ***articulated vehicle***; ***locomotive***.

truck-trailer A commercial vehicle trailer. See ***full trailer***; ***semi-trailer***.

true joint angle Acute angle described by the intersection of the rotational axes of the input and output shafts of a ***universal joint*** and measured in the plane of these axes.

trunk (UK: boot) Rear luggage compartment of a passenger car, internally isolated from the passenger compartment, and with a lifting lid to facilitate access.

trunnion axle Beam axle suspended at its mid-point to facilitate unconstrained movement in the roll, as for example on uneven terrain.

tube (1) The inner tube of a ***tire***. (Informal) (2) A hollow cylinder, usually made from one homogeneous material such as a metal or elastomer. In automotive connotations a tube is usually narrow in relation to its length, and capable of being bent or permanently formed. See also ***hose***; ***pipe***.

tubeless tire Tire in which the air pressure is contained by the ***carcass*** of the tire and the ***rim*** of the wheel, there being no ***inner tube***.

tumblehome Of a vehicle body, inward sloping toward the top in section, from maritime terminology.

tune (1) To adjust the operating variables of an engine (ignition timing, fuel mixture metering, etc.) to give optimum performance. (2) The running condition of an engine with regard to such variables.

tuned absorber Acoustic device for attenuating sound generation or transmission in specific and troublesome frequency bands.

tuned intake pressure charging Increasing the mass of inducted fuel mixture by matching the acoustic resonance of the induction system to engine speed. Also ***tuned intake tube charging***. See also ***harmonic induction engine***; ***ram air induction***.

tungsten halogen lamp Lamp bulb or ***sealed beam*** unit filled with the halogen iodine, which combines with the evaporated tungsten of the filament to form tungsten iodide, which does not deposit an opaque film within the bulb envelope but ionizes redepositing the tungsten on the filament. The glass envelope is normally made of quartz, hence quartz halogen. Also ***quartz iodine lamp***.

turbine (1) A wheel with blades or vanes, driven by a flow of gas, as in a ***turbocharger***. (2) A gas turbine engine. (Informal) (3) The output wheel of a ***torque converter*** or ***fluid flywheel***, which derives its energy of rotation from the circulatory flow of a fluid. See Figure T.3.

turbo-compound engine Exhaust turbocharged ***compression ignition*** engine deriving a substantial part of its power from exhaust gas energy, as for example the gearing of the turbine to the engine output shaft.

turbo-supercharger See ***turbocharger***.

turbocharger Induction pressure charger normally comprising an exhaust gas driven air turbine driving an air compressor. See also ***blown***; ***supercharger***.

turn signal lamp (UK: direction indicator) Signalling element of a turn signal system to indicate intention of driver to turn. Also ***flasher***. (Informal)

turning center Center of the radius of the turn of a vehicle, at which the axes of all wheels ideally coincide. See Figure U.1.

turning circle (1) Radius of the circle about which a vehicle turns when steered. (2) The minimum diameter of the circle within which a vehicle turns when steered at ***full lock***.

turning radius The radius of the ***turning circle***, measured either to the nearest point on the plane of symmetry, or to the ***center of tire contact*** of the wheel describing the largest circle.

turning track Radial width between centers of road contact (centers of contact patches) of innermost and outermost tires of a vehicle when negotiating a turn. In the case of ***dual tires***, center of road contact is taken to be midway between those of individual tires.

turntable (1) The flat bearing surfaces of a ***fifth wheel***. (2) The rotating extension ladder of a fire-fighting vehicle.

twin cam Informal and confusing contraction of ***twin camshaft***.

twin camshaft Arrangement of two parallel camshafts per bank of cylinders in an engine, normally with one operating the intake and the other the exhaust valves. Not necessarily an ***overhead camshaft*** arrangement. Also ***dual camshaft***. See also ***twin overhead camshaft***; ***dual overhead camshaft*** (mainly US usage); ***DOHC***.

twin carburetor Having two (usually balanced) carburetors. Also ***TC***.

twin choke Carburetor having two (usually parallel) ***venturis***. Also ***dual venturi carburetor***. See ***choke***.

twin leading shoe brake Drum brake with two ***leading shoes***.

twin overhead camshaft Arrangement of two overhead camshafts per bank of cylinders in an engine. See also ***twin camshaft***.

twin screw Vehicle with two driven ***tandem axles***. (US informal, from marine terminology)

twist grip Hand control, as for the ***throttle*** of a ***motorcycle***, in which the rotation of a handgrip controls the length of a shrouded cable.

twistlock Manually operated lock for attaching a ***demountable*** to the bolster of the carrying vehicle.

two axle rigid Commercial vehicle with load-carrying ability and two axles, one or both of which may be driven. See also ***rigid truck***. See Figure C.13.

two cycle See ***two stroke***.

two-piece construction Construction of an item in two usually separable pieces, but particularly the construction of a heavy vehicle road wheel in which a detachable flange, such as a spring flange, facilitates the changing of tires. See also ***three-piece construction***.

two-piece wheel Wheel with a detachable rim to facilitate tire replacement. See also ***split wheel***.

two-speed axle Axle with two reduction stages, one of which is a change-speed gear train with a choice of high or low ratios. Also ***double-reduction axle***.

two-spring non-reactive suspension A form of heavy vehicle ***bogie*** suspension in which two inverted ***leaf springs***, one each side of the chassis, are pivoted at mid-length, the axles being constrained at each end.

two-stroke Thermodynamic cycle of spark or compression ignition engine in which the principal operations of induction, compression, power and exhaust stroke take place within one revolution of the engine. See Figure T.11.

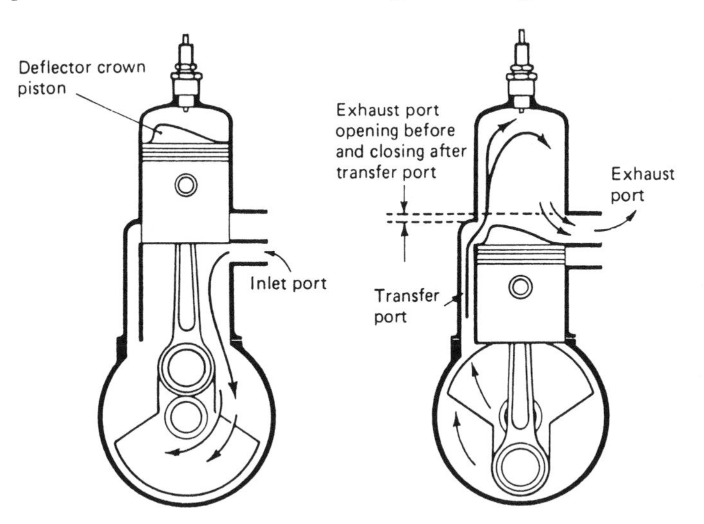

Figure T.11 A crankcase-compression two-stroke engine.

two-tone horn Warning horn that produces two distinct and separate notes.

two-way converter ***Catalytic converter*** containing pellets or honeycomb coated with platinum and palladium. See also ***three-way converter***.

type approval Official ratification of the compliance of a vehicle type with national or international regulations.

tyre See ***tire***.

U

U-engine ***Two-stroke*** engine in which two parallel pistons are driven by one crank, and therefore move in the same direction for most of the cycle. Sometimes called ***split-single***. A variant of this type, the Puch engine, has a Y-shaped connecting rod rather than two separately pivoted rods.

unburned hydrocarbons Unburned or incompletely burned products of engine combustion, particularly as released to the atmosphere as an ***exhaust emission***. Also ***HC***; ***THC***.

underbody (1) The body structure of the underside of a vehicle including floor, wheel wells and stiffening members. (2) The underneath of a vehicle or its effective profile, particularly in an aerodynamic context.

underbody flow The flow of air under a moving vehicle, and the ***aerodynamic*** characteristics thereof.

underbumper apron See ***apron***.

undercarriage The wheels, axles and suspension system of a vehicle, but particularly of a commercial vehicle.

undercoating (UK: underseal) Heavy protective coating applied to the underside of a vehicle to resist corrosion and damage from roadstone impact, and to reduce noise transmitted from under the vehicle.

undercrown The underside of a ***piston crown***. See Figures C.6 and P.2.

underhood (1)The engine compartment. (2) **(UK: bonnet)** The engine and ancillary equipment located under the hood.

underrun protection Transverse structure attached to the rear of a commercial vehicle chassis to prevent a smaller impacting vehicle from running under the chassis.

underscreen See ***undershield***.

underseal See ***undercoating***.

undershield Panelling under an engine compartment, or the whole underside of a car. Also ***underscreen***.

underslung worm transmission Worm and worm wheel final drive in which the worm is set below the worm wheel, giving a low propeller shaft. See also ***overslung worm transmission***.

undersquare engine Engine having a larger ***stroke*** than cylinder ***bore*** diameter.

understeer (1) Response of a vehicle if the ratio of steering wheel angle gradient to overall steering ratio is more than the Ackermann steer angle gradient. (2) Under-response to steering input, as by generation of excessive ***slip angle*** on front wheels. (3) Response of vehicle to steering input characterized by an incremental increase in ***yaw*** rate which necessitates an increase in ***steer angle*** to maintain the intended radius of turn. See Figure U.1.

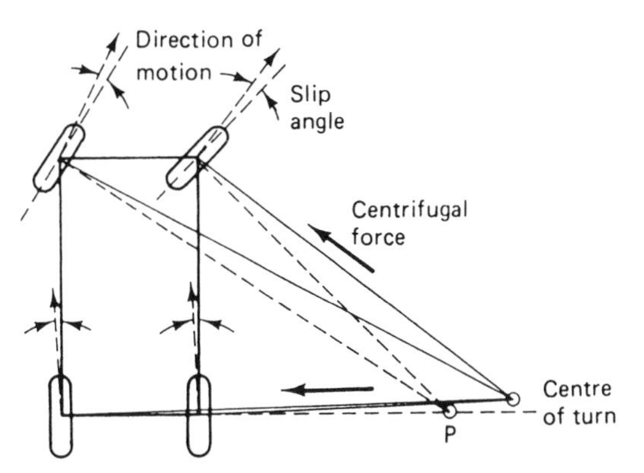

Figure U.1 Diagrammatic representation of understeer, showing the actual center of turn compared with the geometric center of turn that would prevail at low speed.

undertread Reinforcing ***plies*** laid beneath the ***tread*** of a ***tire***.

uniflow scavenging System of scavenging in ***two-stroke*** engines in which the fresh mixture enters the cylinder at one end while the exhaust gases leave through valves or ports at the other.

unit fuel injector Assembly which receives fuel at supply pressure and is then actuated by an engine mechanism to meter and inject the fuel charge into the engine ***combustion chamber***.

unit pump Injection pump containing a single pumping element operated by an engine cam. Mainly used on marine engines.

unitary construction (US: unitized construction) (1) Of a vehicle drive system, the construction of ***engine***, ***clutch*** and ***gearbox*** to form one rigid unit, though enabling each individual item to be detached. (2) Monocoque or chassis-less structure of a vehicle body.

universal joint Rotating shaft coupling that permits angular axial displacement. Some types will also operate with linear misalignment of axes. See also ***Bendix-Tracta joint***; ***Bendix-Weiss joint Cardan shaft***; ***constant-velocity joint***; ***Hooke's joint***; ***Layrub joint***; ***Rzeppa joint***.

Unladen Weight Weight of vehicle (usually a commercial vehicle) normally understood to include driver and passenger and all service ancillaries, but without payload.

unleaded gasoline Gasoline which has no added lead alkyls as octane improvers, but may nevertheless contain a specified very low maximum amount of lead. See also ***lead-free***.

unloader valve Valve that controls or limits air pressure in a pneumatic control system such as a commercial vehicle air brake system, usually by exhausting excess air to atmosphere. Also ***governor valve***. See also ***check valve***; ***pressure protection valve***.

unsprung mass (1) Mass of a vehicle undercarriage below the springs. (2) Mass of vehicle components not carried by the suspension system. In some definitions half the spring and damper mass is included.

unsprung weight Unsprung mass expressed in gravitational units. Loosely used as a synonym for unsprung mass.

updraft carburetor (UK: updraught carburettor) Carburetor in which the inducted air flows upward past a jet, usually centrally located in a venturi. See Figure C.3.

upper beam The main or driving beam of a ***headlamp***.

upper dead center See ***top dead center***.

urban cycle Driving test cycle that simulates driving conditions in a typical urban area.

ute A light commercial or utility vehicle. A ***pick-up truck***. (Australian slang)

V

V-belt Continuous reinforced rubber belt with angled faces that bear on the inner faces of a taper-grooved pulley. Used for driving engine ancillaries such as fan or water pump and, in more robust form, for final drive transmissions. See also ***infinitely variable transmission***; ***Variomatic transmission***.

V-engine Engine of which the cylinders are set at an angle to each other so that the axes form the letter V. The angle is commonly 60 or 90 degrees. See also ***narrow V-engine***.

vacuum advance Mechanical-pneumatic system for advancing ***ignition timing***, using carburetor throat depression as a source of vacuum.

vacuum-assisted brake Hydraulic brake actuated or assisted by atmospheric pressure, usually through mechanical linkages. Also ***air-over-hydraulic brake***.

vacuum brake Conventional brake (for example a ***drum brake***) actuated by atmospheric pressure acting on a piston in a partially exhausted (vacuum) cylinder.

vacuum carburetor Carburetor in which the air slide is controlled by suction or depression from the inlet tract.

vacuum fluorescent display Electronic active (light emitting) display for vehicle instruments such as speedometers, particularly with digital readouts. Also ***VFD***.

vacuum-over-hydraulic brake A vacuum-assisted ***hydraulic brake*** system.

vacuum pump Pump for exhausting air from a cylinder or reservoir, as for operating ***vacuum-assisted brakes***.

vacuum servo Servo system using the differential between atmospheric pressure and a lower pressure source to assist the operation of a mechanical or hydraulic system, such as a brake.

valve (1) Any device for controlling, restricting or interrupting the flow of a fluid. (2) A ***poppet valve*** or mushroom valve for controlling the flow of gases into or out of an engine cylinder. (3) Any form of engine valve. See also ***disc valve***; ***reed valve***; ***rotary valve***; ***sleeve valve***. See Figure V.1.

valve beat-in See ***valve recession***.

valve bounce Bouncing of a ***poppet valve*** on its seat when closing, usually as a result of spring resonance or overspeeding.

valve cap (1) Cap to prevent ingress of moisture and dirt to the inflating valve of a ***tire***. (2) Detachable plug to give access to the valves, particularly of a ***fixed head*** side valve engine.

valve crown See ***valve head***.

valve face The bevelled mating surface of a ***poppet valve***.

valve gear The mechanism that actuates a valve, and in particular the mechanical parts of the valve mechanism from cam to valve. See ***valve train***.

valve guide Tubular insert in engine cylinder head which constrains the concentricity of the ***valve stem*** and seating of the valve. See Figure C.14.

valve head The disc end that performs the sealing operation in a ***poppet valve***. Also ***valve crown***.

valve lifter (1) Engine valve train component that bears directly on the cam. (US) ***Cam follower***. A ***tappet***. (2) Tool for removing valves. (UK)

valve lock Lock nut or other locking item that secures the tappet adjustment of a ***valve train*** once set. Also ***keeper***.

valve overlap Period, usually expressed in degrees of crankshaft rotation, between the opening of ***intake valve*** and the closing of the ***exhaust valve***.

valve recession Accelerated wear and erosion of an engine ***valve seat***, particularly of an engine designed to run on leaded gasoline when running on unleaded gasoline. Also ***valve beat-in***; ***valve sink***.

valve rocker See ***rocker arm***.

valve rotator Mechanism for rotating a ***poppet valve*** while engine is running.

valve seat The annular bevelled surface with which a ***poppet valve*** closes.

valve seat insert Ring-shaped insert of harder or more durable material than that of the cylinder head of an engine.

valve sink See ***valve recession***.

valve spring Spring, usually a coil spring, that returns a ***poppet valve*** to its closed position. See Figure V.1.

valve spring shimmy Resonant vibration of an engine ***valve spring***.

valve stem Narrow cylindrical rod to which valve disc of ***poppet valve*** is attached.

valve stem seal A circular seal, often of a high-temperature elastomer, that acts as a fluid seal to the reciprocating valve stem beyond the valve guide.

valve timing Geometric positions in relation to ***crankshaft*** rotation or other reference at which valves open and close.

valve train The total mechanism from ***camshaft*** to valve of an engine that actuates the lifting and closing of a valve. See Figure V.1.

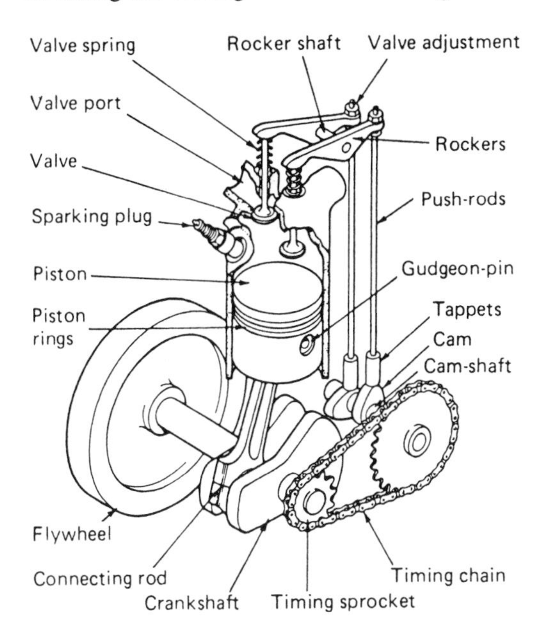

Figure V.1 Valve train of a single-cylinder overhead valve engine.

van Vehicle with fully enclosed body primarily for transportation of freight or for trade use. See also ***box van***; ***demountable***; ***luton***; ***panel body***. See Figure L.5.

vapor lock (UK: vapour lock) Interruption of flow in a piped fluid system (such as a fuel line) resulting from vaporisation of the fluid or entrainment of air.

Vapor Lock Index Index that combines Reid Vapor Pressure and the percentage evaporated at specified temperature of a gasoline. It is a measure of the likelihood of vapor locking.

vapor recovery unit Device for recovering evaporated vapors, as from a gas (petrol) pump.

variable choke carburetor See ***constant depression carburetor***.

variable compression engine Usually a standard engine for laboratory or research use in which the ***compression ratio*** can be varied, particularly while the engine is operating.

variable depression carburetor See ***fixed choke carburetor***.

variable torque dividing On a ***four-wheel drive*** vehicle, proportioning the torque delivered to each axle according to the axle loading. Also ***biasing***.

Variomatic transmission Infinitely variable light car transmission using rubber belts on expanding pulleys, originally produced by van Doorn in the Netherlands.

varnish (1) A clear lacquer. (2) A lacquer-like deposit composed of products of combustion and lubricant breakdown, often occurring on ***piston skirts***. See also ***gum***.

vehicle clearance circle Diameter of the smallest circle which will enclose the outermost points of projection of a vehicle when executing a turn at low speed and full lock.

Vehicle Identification Number (VIN) Number assigned to a vehicle by the manufacturer primarily for registration and identification purposes. It may consist of numerals and letters.

ventilated disc Disc of ***disc brake*** system consisting of two discs separated by ribs to bring about the flow of cooling air between the discs.

ventilation Provision of a free or forced draft of air to and from a vehicle passenger compartment.

venturi Convergent-divergent nozzle which accelerates and lowers static pressure in gases or vapors flowing through it. In a ***carburetor*** the venturi provides the depression in the air flow pressure that causes the fuel to be drawn from its bowl or chamber into the air stream. Also ***choke***. See Figure C.3.

venturi nozzle A venturi-shaped orifice, as for the reduction of low-frequency emissions in an exhaust ***silencing system***.

vertical engine Engine in which the cylinder axes are vertical.

veteran car A car of an early and specified period of construction maintained in original condition.

VFD See ***vacuum fluorescent display***.

VI improver See ***viscosity modifier***.

vibration damper (1) Device utilizing fluid or mechanical friction to reduce vibration of a machine. (2) Rotating or oscillating mechanical device that counteracts vibrations in a machine, and particularly a reciprocating machine. See also ***harmonic damper***.

vintage car Car of a later period than ***veteran***, but particularly one preserved and maintained in original condition. The age of the car is variously specified. See also ***Edwardian car***.

viscosity improver See ***viscosity modifier***.

viscosity index improver Compound added to lubricating oil to improve viscosity index. (Obsolete) See ***viscosity modifier***.

viscosity modifier Polymeric lubricating oil additive which reduces the thinning of oil as it is heated. Once called ***viscosity index improver*** (or ***VI improver***) and ***viscosity improver***.

viscous coupling Friction shaft drive using the property of fluid friction within an ***oil*** or ***thixotropic*** fluid to transmit power.

viscous damping Vibration damping by fluid friction.

void (1) Space between tread elements or ribs of a ***tire***. (2) A gap or space, particularly one which is not normally filled.

volatility Tendency of a substance to evaporate.

voltage droop Undesirable reduction of voltage, particularly of an ***ignition system*** operating at high speeds.

voltage regulator Elecromechanical switch or relay, often with vibrating contacts, for limiting the voltage output from a generator. See also ***control box***.

volumetric efficiency Extent to which the cylinder of an engine is completely filled by the incoming charge following an ***exhaust stroke***. A measure of the ability of an engine to breathe freely. Note that the volumetric efficiency is a ratio of masses, not of volumes. Original definition was the ratio of volume of induced charge at inlet valve temperature and pressure to engine ***swept volume***.

vortex induced drag See ***induced drag***.

vortex pair Revolving wakes of air that follow a vehicle in motion.

vortex stabilizer (1) Vertical, forward projecting plate on the front of a ***semi-trailer***, which obstructs lateral flow through the gap behind the cab in cross-wind conditions and is intended to stabilize the vortex between cab and trailer. (2) Any aerodynamic device intended to promote the orderly formation of trailing vortices, and thereby reducing ***drag***.

vulcanise To react natural rubber and certain ***thermoplastics*** with sulfur or other agent to change its physical properties, and particularly to increase its elasticity and strength.

W

waddle Side-to-side oscillation of a moving vehicle, often as a result of ***suspension*** or ***tire*** damage, or ***lateral runout***.

waffle clutch Clutch of which the friction material is grooved with a criss-cross pattern to aid cooling, giving the appearance of a waffle.

wagon See ***station wagon***.

waist molding Metal or non-metallic strip attached along the ***belt line*** of a vehicle. See ***belt line***.

waistline See ***belt line***.

walk point Beginning of onset of apparent loss of adhesion in cornering, when ***steered wheels*** in an ***understeering*** vehicle appear to walk sideways. (US informal)

walking beam Centrally pivoted beam that supports at each end a swinging shackle of a tandem rear axle ***leaf spring*** suspension. See also ***balance beam***.

wander Directional oscillation of a vehicle or tendency to deviate from the steered direction, for example as a result of incorrect ***steering geometry*** or excessive wear in the steering mechanism.

Wankel engine Rotary engine using trochoidal rotor, after inventor Felix Wankel. See also ***cell swept volume***. See Figure W.1.

warm up Of an engine, the period between starting and reaching a stable running temperature.

wastegate Valve which diverts exhaust gases away from a ***turbocharger*** turbine when charge pressure has reached the requisite figure, or which prevents turbine overspeeding.

water cooled Using water as a heat transfer medium.

water injection Injection of water mist or vapor directly or indirectly into cylinder of IC engine.

water jacket Part of the ***cylinder block*** and head of an engine that encloses the cavity through which cooling water flows. See Figure E.1.

water pump (1) Pump that circulates the water of an engine's cooling system. (2) Any pump for delivering water under pressure, as for screen washing, etc.

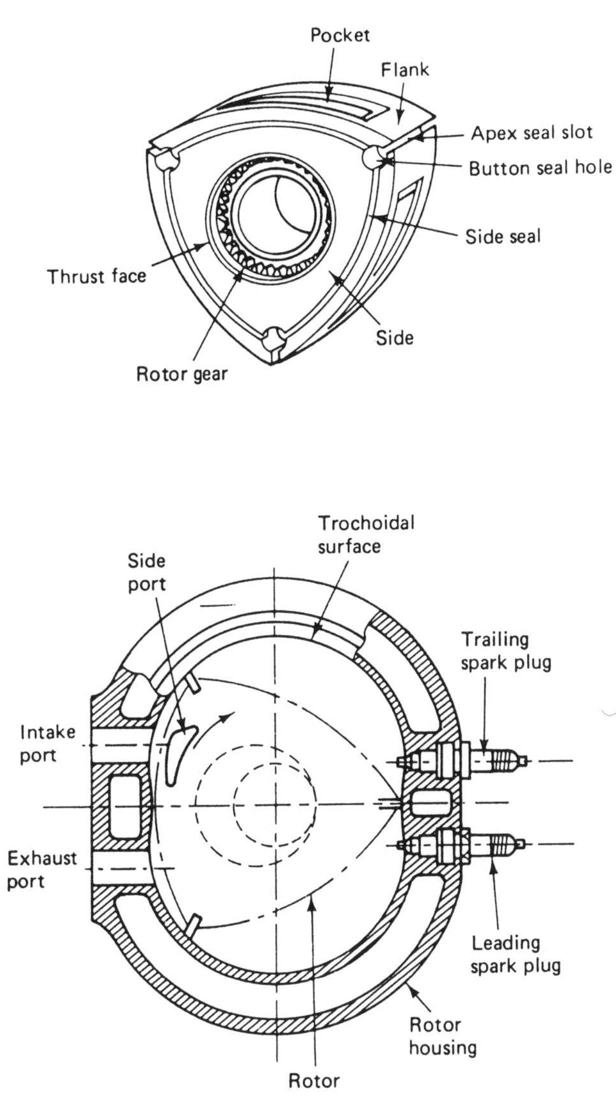

Figure W.1 Rotor and "cylinder" of a Wankel engine.

Watt's linkage Three-element linkage (four if the plane of attachment is included) providing straight line motion of the central element. The three elements usually lie within one plane, the two pivoted outer elements are linked at their inner ends to a short vertically mobile element that follows perpendicular to a line joining the pivots of the outer linkages. Typically used for ***rear axle*** lateral location. See Figure W.2.

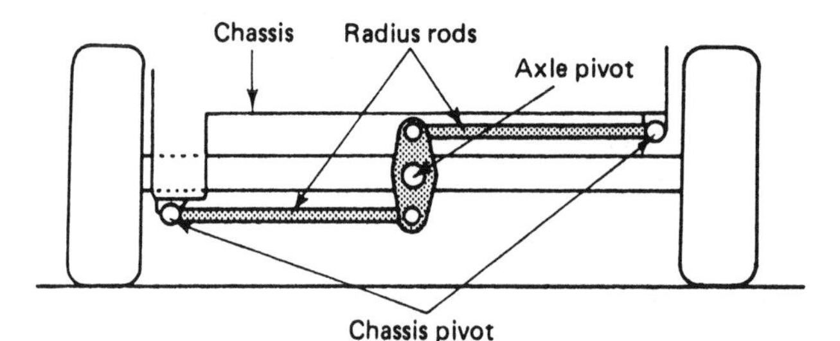

Figure W.2 Watt's linkage lateral location of rear dead axle.

wax High molecular weight ***paraffinic hydrocarbon*** which separates from a lubricating oil or fuel as temperature is lowered. See also ***Cloud Point***.

wax element thermostat Temperature control device normally mounted on or near the radiator to restrict flow of cooling water when the engine is operating at less than optimum temperature.

wax injection Injection of enclosed body components structural members with corrosion-inhibiting wax.

wax pellet thermostat Radiator thermostat in which a wax capsule melts at a predetermined temperature and so releases a valve to allow the escape of steam.

wax plugging Blocking of diesel fuel delivery lines by the formation of wax crystals at low temperature. See also ***Cloud Point***.

waxing The tendency of certain fuels, particularly, diesel fuel, to form wax crystals at low temperatures. (Informal)

weak mixture Air:fuel mixture in which the proportion of air exceeds that for ***stoichiometric*** combustion.

wear adjuster Device to take up lining or pad wear in ***brakes***. See also ***automatic wear adjuster***.

weather cracking Surface cracking, particularly of a ***tire*** surface or other rubber component, often as a result of ozone in the atmosphere. See also ***checking***.

weatherstrip Strip of compressible, flexible material attached around vehicle doors and intended to prevent the ingress of moisture. Also ***weatherstripping***.

web (1) The vertical part of a chassis frame, joining the flanges. (2) The central plane element of a structural joist (I-beam) or channel section.

wedge brake See ***wedge-operated brakes***.

wedge combustion chamber Tapering combustion chamber of overhead valve engine of a type intended to reduce the tendency to detonate. See Figure W.3.

wedge expander Brake expander unit in a ***drum brake*** in which a taper wedge forces apart two pistons or tappets, which bear on the brake shoes.

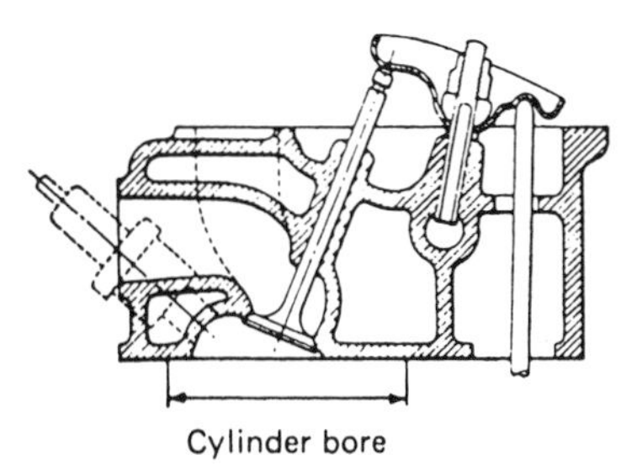

Figure W.3 Wedge combustion chamber with Roesch-type rocker.

wedge lock Sliding wedge of a ***fifth-wheel*** that prevents the ***semi-trailer*** kingpin from disengaging from the hook.

wedge-operated brakes ***Drum brake*** in which shoes are forced apart by a wedge expander.

weight distribution Of a commercial vehicle, the percentage of the total weight distributed to each axle.

weight transfer The effective change of ***axle load*** when a vehicle accelerates or decelerates.

Weiss coupling See ***Bendix Weiss joint***.

Weissach axle Unequal four-bar linkage type ***suspension*** that minimises that steering input due to braking and ***lift-off*** on cornering.

welch plug Disc-shaped pressed metal plug mainly used for closing apertures in engine ***cylinder block*** castings.

well base wheel Wheel with ***one-piece rim*** incorporating at or near its center a well to enable tire ***beads*** to be mounted over the ***rim flanges***.

wet clutch Friction clutch that runs in an oil bath. Also ***oil-immersed clutch***.

wet grip Ability of a ***tire*** to maintain adhesion on wet surfaces.

wet liner (US: wet sleeve) Not strictly a liner as it does not line a cylinder, but is the cylinder itself. (UK informal) See also ***cylinder sleeve***; ***dry liner***; ***wet sleeve cylinder***. See Figure W.4.

wet sleeve Thin-walled hard metal engine cylinder supported at head and crankshaft ends, but in contact with cooling water in between. See Figure W.4.

wet sump An oil reservoir normally attached to the bottom of the ***crankcase***, from which lubricating oil is circulated. (UK)

wet tank Reservoir for water and oil removed from the compressed air system of ***air brakes***.

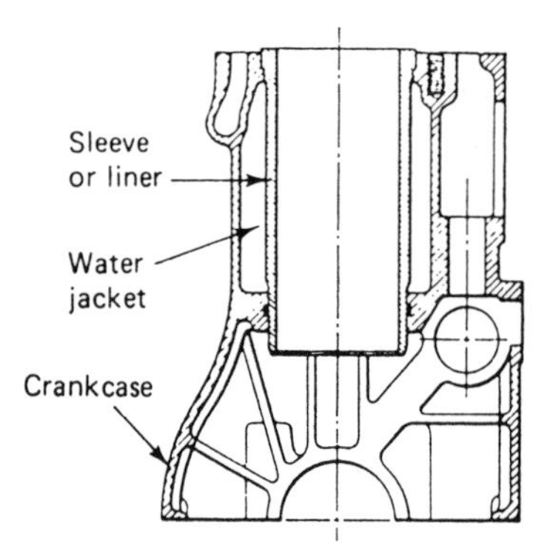

Figure W.4 A wet sleeve or wet liner cylinder.

wets Racing tires for use in wet conditions. (Slang)

Weymann construction Type of vehicle construction in which fabric is stretched over a wooden framework, after carriage-builders of that name. (Obsolete)

wheel (1) The disc assembly on which a vehicle runs. In some definitions the term wheel excludes the tire and hub. See Figure W.5. (2) Assembly of rim and center member or ***spider*** for attachment to an axle. (3) The steering wheel. (Informal)

wheel arch The usually semicircular housing above a roadwheel. See also ***mud-guard***.

wheel axis system Coordinate system used in tire, wheel and vehicle dynamics. Various systems exist originating from the SAE, ISO and other authorities.

wheel base See ***wheelbase***.

wheel cylinder Hydraulic cylinder that forces shoes of a ***drum brake*** apart.

wheel disc The part of a wheel between ***hub*** and ***rim***.

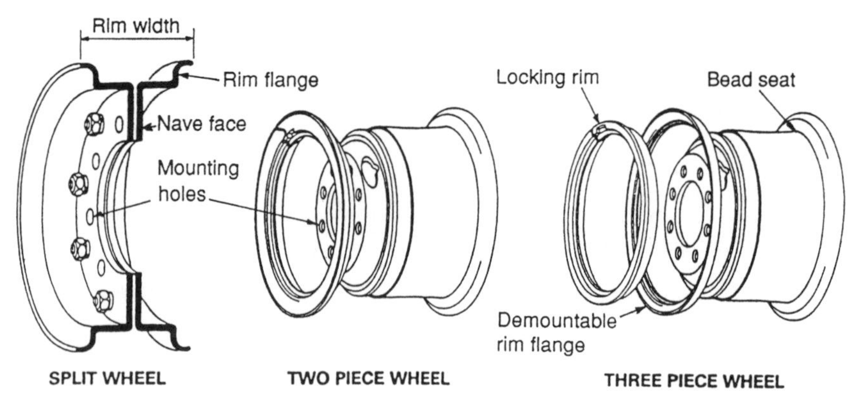

Figure W.5 Three types of truck wheel.

wheel flutter Oscillation of ***steered wheel*** about its ***steering axis***, usually at a frequency greater than that of rotation. See also ***wheel wobble***.

wheel hop Severe oscillation of a wheel in the vertical mode, in which the tire intermittently loses contact with the ground.

wheel nut Nut which fastens onto a stud and thereby attaches a wheel to its hub. See also ***Rudge nut***.

wheel plane Central plane of the wheel, normal to the ***spin axis***. See Figure C.2.

wheel rim The periphery of a wheel on which the tire is mounted. See also ***three-piece construction***; ***two-piece construction***. See Figure W.5.

wheel shimmy See ***shimmy***.

wheel skid Sliding between ***tire*** and road surface.

wheel slip brake control system See ***anti-lock braking system***.

wheel speed sensor Sender device for reading speed of wheel rotation, usually consisting of a toothed pulse-ring and pickup.

wheel spin Rapid slipping rotation of a driven wheel resulting from the application of excessive torque for prevailing friction conditions. See also ***limited slip differential***.

wheel torque Torque applied to the wheel from the vehicle about the ***spin axis***.

wheel well (1) Cavity within a vehicle structure or coachwork to accommodate a wheel. (2) The recessed part of a wheel ***rim***, between the flanges. See also ***nave***.

wheel wobble (1) Oscillation of a wheel at rotational frequency usually resulting from unbalance or misalignment. (2) Self-excited oscillation of ***steered wheels*** about their ***steering axes***, occurring without appreciable tramp. See also ***wheel flutter***.

wheelbase Longitudinal distance between the front and rear wheel axes of a vehicle. See Figure L.3.

wheelfight Rotary disturbance at the ***steering wheel*** produced by forces acting on the ***steered wheels***.

wheelslip (UK: wheelspin) Slipping of driven wheels when friction between tire and road surface is insufficient to react driving torque.

wheelspin See ***wheelslip***.

whiplash Sudden forward and backward movement of the head of an occupant in a vehicle impact. See also ***head restraint***.

whirling Resonance in amplitude of a rotating shaft or flexible component such as a crankshaft. The rotary equivalent to the vibration of the string of a musical instrument.

whirling speed Rotational speed at which resonant whirling takes place, as for example in a ***crankshaft***.

whitemetal Tin-based bearing alloy.

wick carburetor Early form of carburetor, in which induction air gathered fuel vapor from an exposed wick. See also ***gauze carburetor***.

Willan's Line Method Test procedure for estimating or measuring no-load mechanical losses in a diesel engine.

Wilson gearbox A ***pre-selector*** change-speed epicyclic gearbox of specific design.

wind-down See ***axle wind-down***.

wind tunnel Aerodynamic test facility in which air is blown in orderly fashion over an object such as a vehicle or scale model, and instrumented so that quantitive readings of drag, lift and other parameters can be made.

wind-up See ***axle wind-up***.

window recess Distance by which windows of a vehicle are set within the outer body profile, thus exposing structural pillars to the airflow. Also ***window reveal***.

window reveal See ***window recess***.

window gasket Elastomeric seal around a fixed vehicle window.

windscreen (US: windshield) The forward-facing window of a motor vehicle, through which the driver sees.

windscreen angle See ***windshield angle***.

windshield See ***windscreen***.

windshield angle Angle at which the ***windshield*** or ***windscreen*** is set to the horizontal or vertical.

windshield rake Angle of the ***windshield*** or ***windscreen*** to the vertical. See also ***windshield angle***.

windshield wiper (UK: windscreen wiper) Oscillating blade, with flexible rubber blade, for cleaning and removing water from a ***windshield*** or ***windscreen***.

windtone horn Audible warning in which a vibrating diaphragm produces sound by vibrating a column of air in a horn. See also ***fanfare horn***.

wing (1) **(US: fender)** Side panel or formed ***mudguard*** of a vehicle partially shrouding the wheels, to prevent mud or water being thrown up by the wheels. See Figure B.4. (2) See also ***aerofoil***; ***tonneau***. See Figure G.3.

wing mirror (US: side mirror) Rear-view mirror located near either exterior front corner of a vehicle. See also ***driving mirror***.

winglets Small horizontal aerofoils attached to a high-speed vehicle such as a racing car.

winker Flashing indicator or hazard warning lamp. (Slang)

winter front Removable shield, often of fabric, that is attached in front of the ***grille*** to deflect cold winter air from a ***radiator***.

wiper blade Rubber or composition blade of a windscreen or ***windshield wiper***.

wire wheel Wheel of which the hub and rim are linked by wire ***spokes***. (Informal)

wiring harness The complete wiring assembly of a vehicle, particularly when installed as an integrated unit. Also ***wiring loom***.

wiring loom See ***wiring harness***.

wishbone Two-armed or V-shaped frame, mounted in horizontal plane for locating an independently suspended wheel.

withdrawal Disengagement, as of a ***clutch***.

withdrawal bearing See ***clutch release bearing*** (UK); ***throwout bearing*** (US).

withdrawal plate Circular dished clutch element used in certain designs of clutch to disengage the pressure plate from the driven plate. See also ***clutch release lever***; ***throwout sleeve***.

wobble Lateral "out-of-truth" in rotation of a ***wheel flange***.

wobble plate See ***swash plate***.

works trailer Industrial trailer for use on private premises or on public roads with dispensation of the vicinity of the place of work.

worm and lever steering Generic term for any ***steering box*** in which the ***drop arm*** movement is brought about by the action of a follower in a worm. See ***Marles steering gear***; ***worm and nut steering***; ***worm and peg steering***; ***worm and sector steering***.

worm and nut steering Steering gear in which ***drop arm*** action is brought about by movement in translation of a threaded follower on a worm gear, the follower being slotted to engage with a spigot on a crank arm. See Figure W.6.

worm and peg steering Steering gear in which one or two pegs on a crank arm engage with a helical cam. Also called ***peg and cam steering gear***.

worm and roller Mechanism consisting of a worm and a pinion roller in external contact, as used in some types of ***steering gear*** or ***rear axles***. See Figure W.7.

worm and sector steering Steering gear in which a worm, sometimes waisted, acts upon the toothed radiused edge of a sector. A type of ***worm and lever steering*** gear.

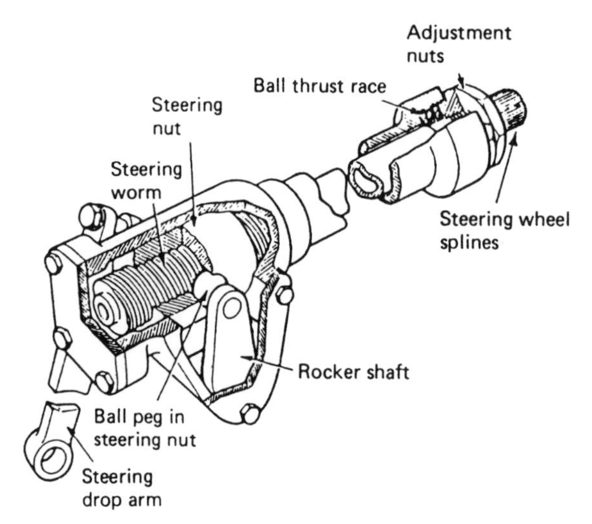

Figure W.6 Worm and nut steering box.

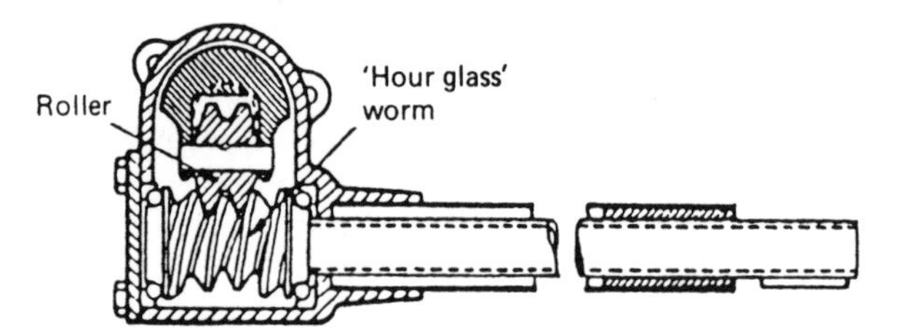

Figure W.7 Hourglass worm and roller steering box.

wrapround Vehicle body element that curves from front or rear to the sides, as for example a wrapround ***windscreen (windshield)***.

wrecker (1) Someone engaged in the recovery of disabled vehicles. (2) A recovery vehicle, particularly one for the recovery of heavy commercial vehicles. (US informal)

write-off (1) A vehicle so severely damaged as to be considered, particularly for insurance purposes, as of zero value or fit for sale only as scrap. (2) To crash or otherwise damage a vehicle so severely as to render it beyond repair.

Y

Y pipe Two branch exhaust manifold connecting the exhausts of a V-engine to form a single exhaust. Also ***breeches pipe***. See Figure E.2.

yaw angle Angle of a vehicle's plane of symmetry or heading to its actual direction of travel or course. Also ***heading angle***.

yaw rate See ***yaw velocity***.

yaw velocity Angular velocity of a vehicle about its vertical axis. Also ***yaw angular velocity***.

yawing moment The moment or applied torque tending to displace a vehicle in the yaw sense.

yawing moment coefficient Non-dimensional measure of the tendency of aerodynamic forces to displace a vehicle in the yaw sense.

yawing rate (1) Frequency of oscillation of a vehicle in yaw. (2) Yaw angular velocity.

yoke A fork-ended component, as of a tow bar or universal joint.

Z

Z-bar See ***labizator***. Zimmermann valve type of crankcase induction disc valve for two-stroke motors.

zinc Metal used mainly in alloy form for intricate castings where strength is not a prime consideration, and as a protective coating for steel. Chemical symbol Zn.

zinc-air battery Battery of high specific energy with negative electrode predominantly carbon and atmospheric oxygen, and a positive electrode of zinc or zinc paste. Various designs exist, some of which are rechargeable. Larger batteries have been used for traction applications.

Appendix

Abbreviations and Acronyms

Readers should refer to main text for definitions of technical terms listed below.

AA	Automobile Association (UK)
AAA	American Automobile Association
AAFI	air-assisted fuel injection
AAMA	American Automobile Manufacturers' Association (formerly MVMA)
AASHO	American Association of State Highway Officials
ABS	anti-lock braking system (also acrylonitrile butydiene styrene)
AC	alternating current (electrical)
ACEA	Association des Constructeurs Europeens (formerly CCMC)
ACL	automatic chassis lubrication
ACORC	Australian Cooperative Octane Requirement Committee
ACRS	air-cushion restraint system
ACS	adsorber-coated substrate (catalyst)
ADR	Accord Européen relatif au transport international des marchandises dangereuses par route
AEA	Automotive Electrical Association (US)
AECD	Auxiliary Emission Control Device
AERA	Automotive Engine Rebuilders Association (US)
A/FR	air/fuel ratio
AHP	accelerator heel point
AIR	air injection reactor
ALU	arithmetic and logical unit (electronics)
ALSC	Automotive Lighting Safety Council (UK)
ANG	adsorbent natural gas (vessel)
ANPRM	Advanced Notice of Proposed Rulemaking (US)
ANSI	American National Standards Institute
API	American Petroleum Institute
APRA	Automotive Parts Rebuilders Association (US)
AQIRP	US Automotive/Oil Air Quality Improvement Research Program
ASME	American Society of Mechanical Engineers
ASR	anti-spin regulation (from German Antriebs Schlupf Regelung)

ASTM	American Society for Testing and Materials
AMT	Automated Manual Transmission (originally an Eaton term)
ATDC	after top dead center
ATF	automatic transmission fluid
ATP	Accord Transport Périssables (French)
AVRO	Association of Vehicle Recovery Operators (UK)
AWD	all-wheel drive
BCC	Bus and Coach Council (UK)
BDC	bottom dead center (of engine stroke)
BEV	barrier equivalent velocity
BHC	burner-heated catalyst
BHP	brake horsepower
BICERI	British Internal Combustion Engine Research Institute (formerly BICERA)
BMEP	brake mean effective pressure
BMV	Bundesministerium für Verkehr (Ministry of Transport) (Germany)
BNA	Bureau des Normes de l'Automobile (France)
BSAu	British Standards Automotive (standards)
BS&W	bottom sediment and wear (of lubricant)
BSFC	brake specific fuel consumption
BSI	British Standards Institution
BTC	British Technical Council (of the Motor and Petroleum Industries)
BTDC	before top dead center (of engine stroke)
BThU	British Thermal Unit (also BTU)
BUDC	before upper dead center (prefer BTDC)
C and U	Construction and Use Regulations (UK)
CAD	computer-aided design
CAE	computer-aided engineering; cab alongside engine
CAFE	corporate average fuel economy
CAG	computer-aided gearshift (Scania)
CAM	computer-aided manufacture
CARB	California Air Resources Board
CASS	copper accelerated acetic salt spray (corrosion test)
CBE	cab behind engine
CCMC	Committee of Common Market Constructors (now ACEA)
CCS	controlled combustion system
CD	capacity discharge (electronic ignition)
CEC	Coordinating European Council (standards)
CEFIC	Conseil Européen des Federations de l'Industrie Chimique (European Council of Federations of the Chemical Industry)
CEN	Comité Européen de Normalisation (European Standards Committee)
CFI	central fuel injection
CFPP	cold filter plugging point (fuels and lubricants)
CFR	Cooperative Fuel Research Committee

CI	compression ignition or Cetane Index
CIT	Chartered Institute of Transport (UK)
C/LO	centerline of occupant
CMVSS	Canadian Motor Vehicle Safety Standard
CNG	compressed natural gas
COE	cab over engine
CONCAWE	Conservation for Clean Air and Water in Europe (organisation)
CO	carbon monoxide
CO_2	carbon dioxide
CORC	Cooperative Octane Requirement Committee (European)
Cp	center of pressure (aerodynamic)
CP	cloud point
CPD	cloud point depressant
CRC	Coordinating Research Council (US)
CR-50	anti-knock mixture of methyl and ethyl lead compounds
CSI	cold start injector
CU	conductivity unit
CV	constant velocity
CVCC	compound vortex controlled combustion
CVS	constant volume sampling (emissions)
CVT	continuously variable transmission
CWT	climatic wind tunnel
DERV	commercial grade fuel for Diesel Engined Road Vehicles (UK)
DI	direct injection (diesel engine)
DBP	drawbar pull
DC	direct current (electrical)
DIN	Deutsches Institut für Normung (German standards authority)
DIPI	deposit induced pre-ignition
DIRSI	deposit induced runaway surface ignition
DIS	distributorless ignition system
DISC	direct injection stratified charge (engine)
DLO	daylight opening
DoD	depth of discharge (of electrical battery)
DON	Distribution Octane Number
DOHC	double overhead camshaft
DOT	Department of Transportation (US); Department of Transport (also DTp) (UK)
DROPS	Demountable Rack Off-loading and Pick-up System
DRACO	Driving Accident Co-ordinating Observer
DRIVE	Dedicated Road Infrastructure for Vehicle Safety in Europe
DTp	Department of Transport (UK)
DTL	diode transistor logic (electronic)
DTT	Department of Trade and Transport (formerly DTp) (UK)
EC	European Community (now EU)

ECC	electronic climate control
ECE	Economic Commission for Europe
ECM	electronic control module
ECT	electronically controlled transmission
ECU	electronic control unit (also European Currency Unit)
ECS	evaporative control system (emissions)
EEC	electronic engine control (also European Economic Community)
EECS	evaporative emission control system
EFEG	European Fuel Experts Group
EGO	exhaust gas oxygen
EGR	exhaust gas recirculation
EHC	electrically heated catalyst
EIN	Engine Identification Number
EP	extreme pressure (lubricant) also end point
EPA	Environmental Protection Agency (US)
EIP	external ignition protection
EMC	electromagnetic compatibility
EMI	electromagnetic interference
EPA	Environmental Protection Agency (US)
EPEFE	European Programme on Emissions, Fuels and Engine Technologies
ESC	electronic spark control
ESO	Lambda Exhaust Gas Oxygen sensor
ESV	experimental safety vehicle
ETBE	ethyl tertiary butyl ether
EtOH	ethanol, ethyl alcohol
EU	European Union
EUDC	Extra-Urban Driving Cycle
EUROPIA	European Petroleum Industries Association
EUROSID	European Side Impact Dummy
FAT	Forschungsvereinigung Auomobiltechnik (Germany)
FBP	Final Boiling Point
FEMA	failure mode and effects analysis
FET	field effect transistor
FEVI	Front-End Volatility Index
FFV	flexible fueled vehicle
FI	fuel injection
FIA	fluorescent indicator adsorption (also Federation Internationale de l'Automobile)
FID	flame ionization detector (analyser)
FID	fuel injector driver (electronic)
FIE	fuel injection equipment
FISITA	Fédération Internationale des Sociétés d'Ingenieurs des Techniques de l'Automobile
FMVSS	Federal Motor Vehicle Safety Standard (US)

FTA	Freight Transport Association (UK)
FTP	Federal Test Procedure (US)
FVLI	Flexi-Volatility Index
FWD	front-wheel drive (also four-wheel drive)
GAWR	gross axle weight rating (US)
GC	gas chromatography
GCW	gross combination weight
GDE	Groupe des Entraves (for dismantling vehicle technical trade barriers) (Europe)
GLC	gas/liquid chromatography
GRP	glass-reinforced plastic
GT	Gran Turismo (car)
GTBA	gasoline grade tertiary butyl alcohol
GTW	gross train weight
GVNTA	goods vehicle national type approval
GVTW	gross vehicle test weight (US)
GVW	gross vehicle weight
GVWR	gross vehicle weight rating
HAZ	heat affected zone
HC	hydrocarbons (collective term)
HCT	hydrocarbon trap
HGO	heavy gas oil
HGV	heavy goods vehicle
HMP	high melting point
HOV	high occupancy vehicle
HSDI	high-speed direct injection (diesel)
HSU	Hartridge Smoke Units
HTA	hydrogenated tallow amine
HUCR	highest useful compression ratio
HVI	high viscosity index (lubricants, etc.)
HWFET	Highway Fuel Economy Test
IAC	Idle Air Control
IBP	initial boiling point
ICC	Interstate Commerce Commission (US)
ICEI	Internal Combustion Engine Institute
IDI	indirect injection (diesel engine)
IFP	Institut Française du Pétrole
IFS	independent front suspension
IGT	Institute of Gas Technology (US)
IHP	indicated horsepower
ILEV	inherently low emission vehicle (EPA definition)
IMEP	indicated mean effective pressure
IMI	Institute of the Motor Industry (UK)
IMechE	Institution of Mechanical Engineers (UK)
IMSA	International Motor Sports Association

IOE	intake over exhaust (valve configuration)
IPA	isopropyl alcohol
IRGA	infrared gas analyser
IRS	independent rear suspension
IRTE	Institute of Road Transport Engineers (UK)
IRT	Institut de Recherche des Transports (France)
IRU	International Road Transport Union
ISC	idle speed control
ISCA	International Show Car Association (US)
ISO	International Organisation for Standardization
JAMA	Japanese Automobile Manufacturers' Association
JCB	JC Banford construction vehicle (mainly European use)
JIC	Joint Industry Committee (US)
JPI	Japanese Petroleum Institute
KD	knocked down (for assembly elsewhere)
KLSA	knock-limited spark advance
LCD	liquid crystal display
LCGO	light catalytic gas oil
LDA	laser doppler anemometry
LDC	lower dead center
LED	light emitting diode
LEV	low emission vehicle (CARB emission standard)
LGO	light gas oil
LNG	liquefied natural gas
LOT	light-off time (catalytic converters)
LPG	liquefied petroleum gas
LTFT	low temperature flow test
LVI	low viscosity index (oils)
MAA	Motor Agents Association (UK)
MAF	mass air flow
MAP	manifold absolute pressure
MAS	maximum axle spacing
MAT	manifold air temperature
MAW	maximum axle weight
MBN	Motor Octane Blending Number
MDFI	middle distillate flow improver
MeOH	methanol, methyl alcohol
MEP	mean effective pressure
MIAW	maximum intermediate axle weight
MIRA	Motor Industry Research Association (UK)
MMT	methylcyclopentadiene manganese tricarbonyl
MON	Motor Octane Number
MOS	metal oxide semiconductor
MoT	Ministry of Transport (former UK ministry)
MPI	multi-point injection

MPV	multi-purpose vehicle (also multi-purpose passenger vehicle)
MRT	mean repair time
MS	Military Standard (US)
MTBE	methyl tertiary butyl ether (fuel additive)
MTBF	mean time between failure
MVEG	Motor Vehicle Emissions Group (EC)
MVI	medium viscosity index (oils)
MVMA	Motor Vehicle Manufacturers Association (now AAMA)
NDIR	non-dispersive infrared
NDUV	non-dispersive ultraviolet
NGV	natural gas vehicle
NHTSA	National Highway Traffic Safety Administration (US)
NIST	National Institute of Standards and Technology (US)
NO_x	oxides of nitrogen (collectively)
NPRA	National Petroleum Refiners Association
NSSN	National Standards System Network (US)
NTA	National Type Approval (UK)
NVH	noise, vibration and harshness
OEM	original equipment manufacturer
OFP	ozone forming potential
OGV	overhead valve
OHC	overhead camshaft
ORI	octane requirement increase
ORV	off-road vehicle (US)
OTA	Office of Technology Assessment (US)
OTC	Ozone Transportation Commission (US)
PAH	polyaromatic hydrocarbon, polycyclic aromatic hydrocarbon
PC	polycarbonate (high-strength polymer)
PCV	positive crankcase ventilation
PFI	port fuel injection
PFR	primary reference fuel
PIN	product identification number
PLA	programmable logic array (electronic)
PNA	polynuclear aromatics
POM	acetal resin plastics, polycyclic organic matter
PP	polypropylene (thermoplastic)
PPM	parts per million (also ppm)
PPO	polyphenylene oxide
PR	ply rating (of tire)
PRF	primary reference fuel
PROMETHEUS	Programme for European Traffic with the Highest Efficiency and Unmatched Safety
PR	ply rating (of tires)
PS	Pferdstärke (German horsepower)
PSU	polysulfone (thermoplastic)

PSV	public service vehicle
PSD	power spectral density (acoustic and vibration analysis)
PTFE	polytetrafluoroethylene (low-friction polymer)
PTO	power take-off
PTS	Permanent Threshold Shift (of hearing acuity)
QFD	quality function development
QMS	quality monitoring system
RAC	Royal Automobile Club (UK)
RBN	Research Blending Number
RHA	Road Haulage Association (UK)
RON	Research Octane Number
ROPS	roll-over protective structure
RSI	runaway surface ignition
RUFIT	Rational Utilization of Fuels in Transport
RV	recreational vehicle
RVI	Recreational Vehicle Institute (US)
RAC	Royal Automobile Club (UK)
RTL	resistor-transistor logic (electronic)
SAE	Society of Automotive Engineers (US)
SAMOVAR	Safety Assessment Monitoring On-Vehicle with Automatic Recording
SAMT	Semi-Automatic Manual Transmission (Eaton)
SCF	standard cubic feet (of gas at NTP)
SE	specific energy
SFC	specific fuel consumption
SFI	sequential fuel injection
SFUDS	Simplified Federal Urban Driving Schedule
SG	spheroidal graphite (iron)
SHED	Sealed Housing for Evaporative Determination
SHP	shaft horsepower (rare in automobile context)
SHPD	super high performance diesel (lubricant)
SI	spark ignition (also Système Internationale d'Unites)
SIR	supplemental inflatable restraint
SMMH	scheduled maintenance man-hours
SMMT	Society of Motor Manufacturers and Traders (UK)
SNAP	Significant New Alternative Policy (US official program)
SNG	synthetic natural gas
SOHC	single overhead camshaft
TAME	tertiary amyl methyl ether
TBA	tertiary butyl alcohol
TBI	throttle body injection
TCP	tri-cresyl phosphate (fuel additive)
TC	twin carburetor (also twin camshaft)
TDC	top dead center (of engine stroke)
TDR	total driving resistance

TEL	tetra ethyl lead (additive)
THC	total hydrocarbon (emissions)
TIR	Transports Internationaux Routières (European)
TLEV	transitional low emission vehicle (CARB emission standard)
TML	tetramethyl lead (additive)
TOFC	trailer on flat-car (transportation)
TPV	thermoplastic vulcanizate
TRA	Tire and Rim Association (US)
TRL	Transport Research Laboratory (UK)
TTC	tire track circle
TTL	transistor-transistor logic
TTS	temporary threshold shift (of hearing acuity)
TUV	Technischer Überwachungsverein (technical inspectorate)
TVO	tractor vaporizing oil (fuel)
UDDS	Urban Dynamometer Schedule
UJ	universal joint
ULEV	ultra-low-emission vehicle (California)
ULW	Unladen Weight (of commercial vehicle)
UKPIA	United Kingdom Petroleum Industry Association
USABC	United States Advanced Battery Consortium
UTAC	Union Technique de l'Automobile, du Motocycle et du Cycle (France)
VCA	Vehicle Certification Agency (UK)
VCM	vehicle condition monitoring
VCR	variable compression ratio
VDI	Verein Deutsche Industrie (Germany)
VEC	vehicle emission configuration
VFD	vacuum fluorescent display
VI	viscosity index
VIN	vehicle identification number
VHVI	very high viscosity index (oils)
VLI	Vapor Lock Index
VOC	volatile organic compound(s)
VSE	Vehicle Standards and Engineering Division (DOT)
VTAC	Vehicle Type Approval Center
VVA	variable valve actuation
WAP	wax appearance point
WASA	wax anti-settling agent
WCM	wax crystal modifier
WMI	world manufacturer identifier
WOT	wide open throttle
WVTA	whole vehicle type-approval (European)
VWTA	whole vehicle type approval
ZEV	zero emission vehicle